# Frost and Grass

### A tour of the invisible

Written and Illustrated by

Sigmond Axel

# Contents

# Introduction

Frost and Grass have been interacting invisibly for eons. Cool and cold season grasses sense when days shorten and temperature drop slowly. The processes driving Frost follow a consistent yet tricky dance. Add The third player the Earths complex surface and let the games begin.

In other words condensation like sweat on a cold drink happens. A cuze or insulator can slow the process. Equalization of temperature will happen eventually. This is an invisible process describes a natural equilibrium the basic force in our shared environment.

Our focus here is on understanding how to reasonably estimate when a Frost event will occur. And to limit damage to Grass by players and people. To do this we dive deep into Frost and its mechanisms. All nearly invisible processes. Then we explore how Grass has adapted to this natural force.

In addition to some new perspectives and armed some modern tools like weather apps and hand held infrared thermometers some simple heuristics for determining a Frost event are revealed. Also by looking closely at the error with modern weather systems and weather equipment standards on a large scale shows why it is difficult for these systems to accurately forecast local Frost events.

Several heuristics are used to estimate the occurrence of a Frost event and when to stay off grass. Of utmost importance is how recovery and thawing of Frost works so a quality time estimate can be made to enable  people to use a Grass surface.

With Grass we will explore why Frost forms on Grass. Then a tour of some of the particulars of what Grass adds to the Frost event. Of key interest here is how and when to avoid damage to Grass. Then add the difference between frozen Grass and Grass with Frost. Finally we conclude by bringing Frost and Grass sections together.

This books sole intent is to assist you in your own discovery of the Frost prediction and recovery process. This can be used and is not intended to be used for the express protection of expensive assets. This has simply been a fantastic personal journey worth sharing.

Sigmond Axel 2025

# The Boundary Layer

Our atmosphere has been described as the same thickness as a skin of an apple over our planet. Inside this skin the weather systems swirl with highs and lows moving Eastward all the while the Sun rotates Westward.

Most of the space in the atmosphere is taken up with gasses, water vapor and particulates. Our topic is Frost and Grass. The Boundary layer or the lowest most atmospheric layer this is where we live. Dew or condensation and frost are formed on just about any surface with the right conditions. Dew and Frost are visible yet we cannot see many of the interactions that precede it. We can deduce that they exist from some simple experiments.

*...our story of Frost and Grass starts here*

The Boundary Layer as defined by atmospheric scientists is a turbulent and ever changing place where sky meets the land. Much less stable and more chaotic than the higher atmospheric layers. Which makes it a place where mathematics and other estimation tools do not apply. The many different complex interactions at ground level produce chaotic and unpredictable events. The views and perspectives are many. For example a quality perspective of this environment would include all the living things at ground level. Plants produce literally tonnes of Oxygen O2 and Carbon Dioxide CO2 per day and per night. So do humans. The beginning of our twisted story of Frost and Grass starts here with chaos and living things coexisting at the Boundary level. Our goal is to hunt the illusive Frost event.

*The Boundary layer is an open system.*

Most of our understanding of gases and atmospheric rules come from closed systems tested under controlled situations. Very useful tools and concepts have been demonstrated and calculated under controlled conditions. Molecular weights of gasses, and how they function under pressure is very useful here.

There are great theoretical thermodynamic tools to estimate energy and how it transfers. The caveat is the Boundary layer is an open system. Meaning processes are not controlled like lab experiments. They are free and seemingly unconstrained. Equilibrium and natures balancing act is one thing that applies in the Boundary Layer.

In reality the invisible action on many levels of a push-me pull-you action on the individual molecule level, group molecule level and temperature and pressure. To be accurate unknown forces too. All play a role.

In the Boundary layer there are many different varying conditions that can create strange condensation conditions and also Frost events. The basic starting point to form Frost requires the transition from water vapor or condensation to ice which we will explore in detail.

The term turbulent is an understatement for the Boundary layer. The play between equilibrium and chaos is the main driver of the turbulence. A basic understanding of this interaction or play in open systems tends to equalize in strange ways.

The Boundary Layer is the bottom layer of at least five 5 layers of atmosphere. Maintaining order or just the want for order in a chaotic environment means there is a lot of playing going on. The varying terrain from absolute flat plains to oceans shore lines and abruptly rising mountain regions to hilly uneven terrain defines turbulence close the the Earth's surface.

*Wind distributes and equalizes heat and cold*

There is evidence that the Boundary layer shrinks in the night and expands during the day. All of this is debatable and variable depending on location on the Earth and surrounding terrain shape and influences. It can be unclear what determines the boundary between the lowest level and the next as this may change at night or be influenced by sometimes unknown factors. At times it is one big atmosphere at other times it demonstrates some form of separation. This is just the lowest part. This is the part we are interested in. The actual thin line where Earth meets atmosphere that is where Frost is found. This thin layer is what we are interested in.

## Estimation Error

In general most things in open systems are relatively defined through averages and estimations all with loosely defined error margins. Error margins will be important throughout our journey. To define error in this context is to say error is the difference between any estimation or derivation and the real world.

Our goal here is to get close to a Frost event. This will be like shooting an arrow into the bullseye on a windy day. The first arrow may hit it. The second arrow may be a little off and still close and on target. Error calculations on many atmospheric experiments have been very enlightening and very applicable here. To estimate a complex event like Frost the standard rules do not apply. The mathematical tools need good numbers which are sometimes are hard to find in the time needed. As the Frost event can be instantaneous and literally roll across the landscape.

*a chaotic environment means there is a lot of playing going on*

The World Meteorological Organization WMO [wmo.int] standardizes weather instrument and reporting standards. This means that standardized weather reporting stations are all set at 1.25 meters off the ground in relatively open areas. There is very good reason for this. To limit the chaos and ground interaction with the weather instruments.

Why instruments are 1.25 meters above the surface is basically to reduce the errors. The goal is have the same instrument accuracy standards across all reporting weather stations then theoretically the picture should be pretty accurate. This is clearly not the case. There is builtin variability and error in the Boundary layer. This is why error is important.

An example, if 100 weather stations report the same temperature across a wide area then the accuracy improves. Add just a little variance across say 20 weather stations/data points and the accuracy falls apart and the variance increases. Using the same math with 20 erroneous points can change a standard error margin or level of accuracy.

Not only does the error margin change daily it can improve and degrade randomly and eventually wobble to an average value. When trying to find a specific condition at a specific time like a Frost event the error margins do not help. They make the whole estimation process even more fuzzy.

*Water and its phase properties help us understand Dew and Frost.*

What is more important is the instrument error rates. Five different instruments that measure temperature, humidity, wind, rain, and sometimes Sun all have different error rates. The locations and height of weather stations above the ground can vary widely and there is no really ideal location. Weather stations close to important sites like airports are kept calibrated and checked often for accuracy. What this means is that your weather app accuracy depends on where you are, what weather stations the app is using and many other unknown factors.

The error rates may vary from day to day. Large data sets smooth out some of the error over large areas using algorithms. The contrast and difficulty is that predicting a Frost event requires accurate measurements at the ground level not at 1.25 meters above the ground.

Most large scale weather applications use satellite data and this has its own built in error. The real trouble is that the satellite has a hard time determining actual ground temperature because it will average out the height of the terrain. The error rate for the satellite can really be like having a weather station 25 meters in the air above ground level. The estimations may be very good on flat or consistent terrain. Large scale terrain changes can introduce all kinds of strange error.

Each pass of the satellite creates a data snap shot meaning as the satellite pass over an area that data is only good for that point in time. This is why there are smooth looking transitions on some weather apps and videos created similar to the concept of frames per second in a video. Frost is a very specific instantaneous event and forms when the conditions are just right. All this data is useful. It just has error when dealing with an instant variable event like Frost.

Heuristics or a set of known conditions can help predict a successful Frost event. Add in modern weather applications and local measurement tools like IR thermometers to measure actual ground temperature and estimations can get close. Edge cases on Frost are tricky. To summarize there are layered error margins on error margins that have to be understood and the heuristics help to logically get to a reasonable conclusion reducing the error. Still not perfect yet close.

*Many weather stations are better than one weather station for predicting weather.*

To explore error a little more land based weather station weather sensors need some exploration. Sensors like humidity and temperature typically sense the world around them and then translate that into an adjusted or calibrated output. This is an error correcting curve. The calibrated curve takes a real atmospheric reading and maps it to a calibrated value. Each instruments curve and sensor can be different. Most of the time this makes for a nice accurate translation within a specified error margin.

Sensor hardware can still slide out of calibration over time. Some fancier equipment will use an error correcting algorithm to make the accuracy last longer yet this approach can introduce calculation error and re-calibration is still necessary. The end result is data that is just a little off from the actual. The key here is that weather sensors are all in a state of degradation and constantly separating from the real readings in complex and dynamic ways. There is always a variable error in weather data.

A weather data set can have thousands of weather stations. A large data set is needed for some things like the boundary of a high and low pressure systems over a large area. The error is reduced with more than one ground weather station and additional satellite passes or different types of weather data mixed together.

For temperature the error in a large data set can be as high as plus or minus 4C or 5.4F off the real temperature in some areas yet not all of them. Welcome the Boundary layer. That error is just temperature. Temperature and Relative Humidity RH are used to calculate Dew point or the point of water condensation. This is an example of layered error. A error containing temperature is used to calculate a Dew point. The error is transferred to the Dew point.

*Accurate prediction of Frost is helped by direct measurements*

Relative humidity RH instruments are notoriously inaccurate at higher humidity readings. Relative humidity is the measure of water vapor relative to the maximum water vapor or Dew Point at that atmospheric or barometric pressure and temperature. The temperature has to be included when referring to Relative Humidity RH. Dew point is always 100 percent humidity at a specific temperature and atmospheric pressure. Most of the time Relative humidity is calculated from temperature which adds the standard error of plus or minus two 2 percent %.

These human designed measurements of our atmosphere give us a small window into weather at the Boundary layer. The more we look into weather in the Boundary layer the more complex and impossible it seems. With all the inconsistencies and variance there is a top limit for the error variance.

Temperature measurement is the most accurate. If the humidity sensor is off by two 2 percent % and the temperature that was used to calculate the humidity was off by on 1 degree for what ever reason. A Dew point calculation on a small sensor may not use floating point math which introduces another error. The Dew point that is presented may be way off from the actual. There are several different ways to calculate relative humidity RH. All error prone. The accuracy of the Dew point depends the error of the previous instruments and calculation error. When it comes to Frost these errors can be really tricky to interpret. With the right heuristics and some local testing and adding in the local terrain anomalies the errors can slowly be eliminated. All of a sudden when you can see your breath the conditions are perfect for condensation.

*When you can see your breath there is condensation in the air.*

Error margins in temperature, humidity and Dew point play an important role in our estimation of Frost and Dew. New understandings of the depth of error for your extended surrounding area and your local terrain may help with your predictions. A higher accuracy solution is to use a preferred calibrated local weather station. Or to use a weather application with the addition of a local micro climate setting.

Quality calibrated weather stations are near airports or other important sites. These are maintained and calibrated regularly. Consistently using the same app and adjusting its values with a known reference like either a hand held weather instrument instrument or an IR thermometer and a separate device to measure water vapor and do the calculations yourself or use a chart like the air psychronometric chart in appendix 1.

*The psychronometric chart is great for validating Dew Point and Relative Humidity (Appendix 1).*

By acknowledging the error factor for temperature, humidity and Dew point there is a good chance of getting a fairly close and accurate estimate of what is really going on for your area.

The error is very important in the context of an open system and especially in the turbulent and constantly changing Boundary Layer. Frost can be elusive. With modern weather apps and systems weather models can get close to an accurate estimate of a probability of Frost. Accuracy is not consistent. One possible reason beyond the instrumentation and data errors is the shifty nature of the Boundary layer.

By drilling down on what is known about the physical attributes of water and a little atmospheric physics and focusing on events some simple heuristics can be deduced to get closer to a Frost event. With some field or local adjustments eventually get to a level of certainty that is not perfect yet close. This can be achieved with some data logging of your own and add some local knowledge of your area and predicting Frost and Dew accuracy can be refined over time.

To really understand and get close to a Frost event requires information from the day or days before and to get better at predicting future Frost events looking at the day or days after. In other words we will need some information on the day before that may not be available if not aware of the mechanics of a Frost event.

Local terrain can add all kinds of different pressure and temperature fluctuations to an area. Just looking at say weather app data on an early morning before a potential Frost event may not provide all that is need to make an accurate prediction.

Frost only occurs where air gets cold enough. Looking at the seasonal day length reduction and the dropping day time temperature in the Fall and Spring indicate the potential for a Frost event. For example, Dew events have been happening all Summer then in the Fall season they increase and can be and indicator of an upcoming Frost event. We will explore that in detail in later chapters. Understanding Dew and how it is formed is integral in understanding Frost.

In some mountains ranges throughout the world frost happens all year long depending on conditions and position on the Earth. For most areas Frost is a seasonal occurrence. For the Northern hemisphere Frost starts in Early Fall and ends in late Spring and opposite in the Southern Hemisphere. Frost can happen in and around the Equator at higher elevations and on mountain Ranges or high plateaus where the right combination of near freezing temperatures and humidity can trigger a Frost event. Abnormal atmospheric conditions can spoof and playfully frustrate Frost prediction. There are Frost events that occur after Sunrise with temperatures above freezing. There can be just Frost in the high grass and not on the lower grass fields. This unpredictable nature of Dew and Frost can be frustrating at first and can become fun and mysterious with a little understanding. This is the dance of Frost and Grass in the Boundary layer.

### Condensation : Glass table top example.

*Outside Glass table top. An outside glass table top in the Sun collects heat during the day the glass collects heat. As the glass top cools during the night more and more water collects on the surface in proportion to the heat released. Once the temperature of the air is cooler than the table top and the temperature of the air just above the table top hits the Dew point water collects right at the surface of the table and will continue until the heat is equalized with the temperature of the table. It is not uncommon to see more water on the tables in the Sun than tables in the shade. This effect is the opposite of the condensation on a cold drink on a Summer day.*

Condensation in the form of water droplets on a cold glass or can is a most obvious display of a weather process. The air around the table where this coldness is cannot hold any more water at that cold temperature and humidity so it collects and forms water droplets. With the table condensation example notice that there is small amount of water clinging to the underneath of the table because heat rises.

This is how condensation forms on grass except that grass has small leaves that tend to hold up the small droplets of water. The small droplets of water distributed across the grass represent how the grass disperses or exchanges heat just like the cold table top glass or can.

From lab experiments it is known that to condense water takes longer than to freeze water. With Frost the smaller droplets freeze faster and thaw faster. Just about the same amount of energy theoretically is used to thaw water as is to freeze water. In the case of Frost on Grass the thaw can be calculated and can be less than 20 minutes with the right temperature and a warming wind.

There is error in everything when dealing with Frost. The quick spontaneous and variable nature of Frost has traditionally been very illusive. By explaining the error in modern prediction systems and revealing that here maybe many layered errors can form a path to understanding the Frost event. To expect error when dealing with Frost is the whole point of this chapter. The intention is not to discount the great work by atmospheric scientists as these modern systems are really incredible tools that can help in getting close to predicting a Frost event.

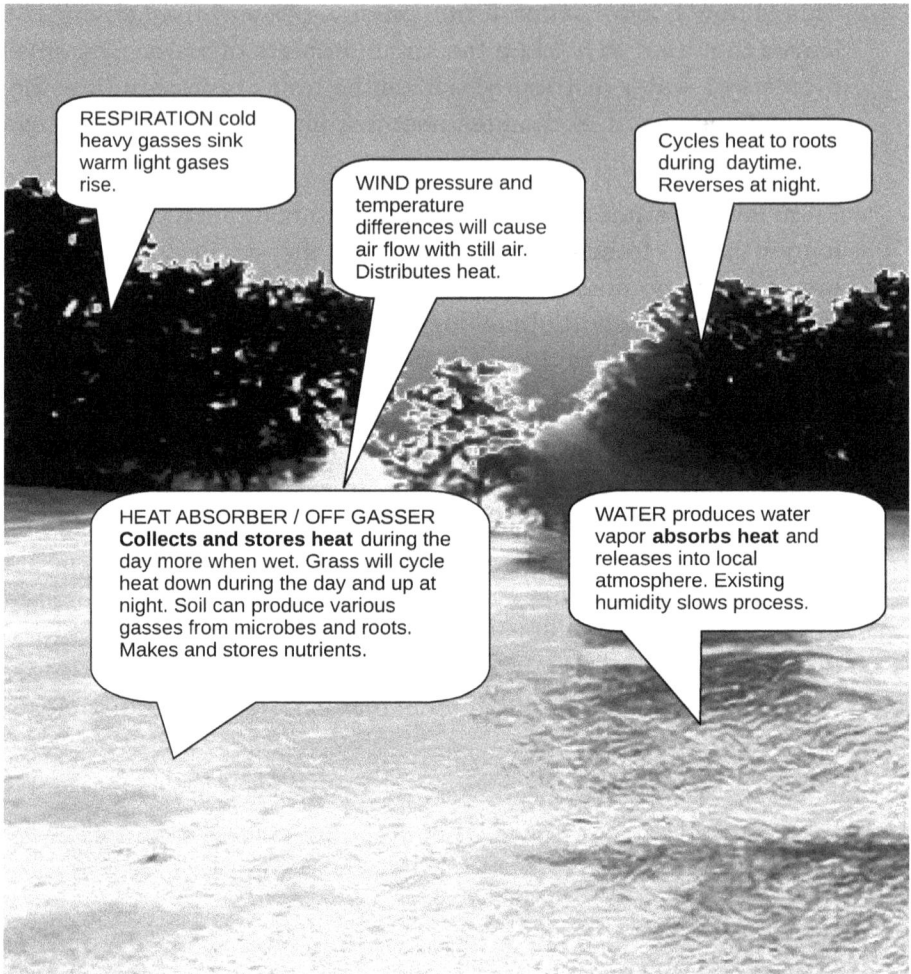

Figure 1: Heat distribution and equilibrium in the Boundary Layer or ground level.

The great water cycle has been going on for eons changing water vapor into liquid water and then to ice and back again reversing from ice to liquid then to water vapor. Water vapor holds heat then rises and creates rain and snow which falls to the ground only to melt return as water vapor to do it all again.

# Water Phases

We know that this process is free of charge. Meaning the energy it takes to go through the phases of water is the same going forward as backward. From ice to water and water to ice; Frost. From water vapor back to water back to water vapor; Dew. Then sometimes to ice cold enough. Frost on grass is nothing more than small ice droplets distributed over a wide grassy area. Typically formed on the leaves of grass or on other ground cover near the ground surface level. This does not discount that Frost could also form on the tops of trees as well.

When looking up at the moon and stars on a clear night sometimes a haze is visible other days clear. We do not see the end of one atmospheric layer and the beginning of another to us it is just sky. The layers are concepts we use to divide the sky and the processes at work there into distinct segments.

Since this is not a laboratory the atmosphere is what is called an open system. In open systems anything goes. Anything can happen. Error margins can be illusive and troublesome. What was thought to be predictable typically needs to be tested and validated with local instruments and local data in combination with the weather application or app data for your local area. The description of the error factors in the previous section is a testament to the notorious nature of the Boundary layer.

Water phases explain how water H20 molecules behave in the atmosphere. Water transitions from a solid as ice to liquid as water to gas as water vapor. Solid ice turns solid at a specific temperature and pressure. When ice transforms into water it takes roughly the same energy almost every time it transforms. It still takes energy.

Energy we are interested here is in heating and cooling. There are many complex forms of energy and energy transfer in the Boundary layer as we shall see. Starting with molecule level there are chemical interactions. One example of this is ozone interacting with water on the ground to create an almost instant Frost. Also not to be out done by massive interactions of large areas of water vapor or cloud fronts interacting with warm air with or with out water vapor. Then there are the pollutants and green house gasses. For our context the way water transforms into water vapor it takes a very specific amount of energy to do so is of special interest.

Each water phase transition takes energy. This energy comes from gravity where light air rises. Denser colder air sinks. More importantly the movement of energy from hot to cold and back to cold from hot.

There is a very curious triple point for water also known as a zero point. Just before freezing at this special temperature water can transition directly from water vapor to Frost or ice. The triple point or zero point facilitates the phase movement in both directions depending on the movement or rising and falling of the temperature. Most ideal gases have to have a set pressure for this triple point or zero point transitions to occur. These energy free transitions are well known for pure types of gases. Water is not an ideal gas and the pressure at the zero point can be has to be in a short temperature range in and around 0C or 32F.

*Saturated air can only hold so much water vapor*

There is another phase transition when air becomes over saturated with water vapor. Saturated air can only hold so much water vapor at a specific temperature and pressure. Water will precipitate out of air that cannot hold any more water vapor when the temperature is dropped. Its like certain type of "air" or ground air will hold as much water as it can at a temperature or pressure.

When pressure or temperature changes water condenses out of water laden air. See the glass table example above or relax with a cold drink  on a hot summer day and watch the condensation pull the cold out of the can or glass and leaving the surface coated with water droplets. The frosted mug in the freezer is the same principle. Warm mugs put into the freezer make a frosty mug. Cold mugs not so much.

When it comes to talking about Frost and Dew on or over Grass every random detail can help with a good Frost forecast. Frost mostly happens on clear calm nights. A clear night with an invisible or very thin cloud layer can influence whether a Frost event happens or not. A thin cloud layer may not visible with our eyes at night. Yet a radar frequency may see it clearly. If this cloud layer thickens and traps heat from escaping this will limit the release of heat from the ground surface.

All the conditions for a Frost event may seem to be evident and in reality Frost does not happen. Using the mug example it was cold to begin with so not much Frost. The grass may freeze instead of Frost if the temperatures drop below freezing really quickly. This is evident with a slick coating of ice on the grass. Its not frosty its clear. Frozen grass is prone to damage. Caution should be taken. Some attention to detail is needed to really get good at predicting the Frost event.

*Water phases transition from water vapor (gas) to water (liquid) to ice (solid) and back again*

Daily Frost event times are early evening, night, morning before sunrise and after sunrise. The seasonal event times are in the Spring and Fall seasons when nights are longer than days. Also it depends on location and altitude above sea level. Winter with its short days and long nights in the Northern Hemisphere and opposite in the Southern Hemisphere is always vulnerable to Frost at the right latitudes. On the equator high mountains can Frost as the same mechanisms in the water cycle are universal. Frost is one aspect of the water cycle and it is universal or world wide and happens predictably with the right conditions.

At the equator Frost can happen at an elevation all year round with the right circumstances even though the night and day length are nearly the same. For Frost to form water has to go through several states from water vapor, to water or liquid to ice. Frost in our definition here is small droplets of ice on leaves of grass.

# Frost Cycle

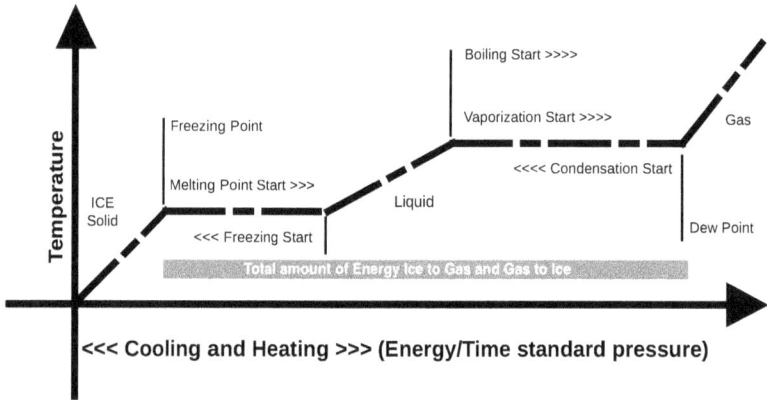

*Figure 2: Water energy phases bidirectional diagram. Ice to liquid to Gas and Gas to liquid to solid ice. At standard pressure (1 atm) sea level. The total amount of energy does not change to convert water vapor to Ice and Ice to water vapor.*

Figure 2 is a graph or demonstration of how water transforms from ice to water vapor and water vapor to ice at standard pressure 1 atm at sea level. The amount of energy is roughly the same to form Dew into Frost (moving from Right to Left figure 2) and from Frost to Dew (moving Right figure2). One atmosphere of pressure is constant in Figure 2. The Boundary layer acts as if chaos rules. Figure 2 at one atmosphere pressure is one thin slice of reality.

To explain Figure 2. Starting at the left figure 2 with ice and moving right. Ice at a temperature of 0C or 32F. Ice is solid at this cold temperature and as the X axis indicated there is little or no heat is added when ice is Frozen. Moving right adding skies the ice starts to warm. At some point after enough heat the ice starts to melt. From this melting point more heat is required to transform all the ice and water to liquid. Now the liquid water absorbs heat. And continues to absorb heat until vaporization starts and finishes as all the water has turned to vapor.

Starting from the far right figure 2 now equalizing the heat going left. Water vapor at this point removes enough heat to get to the Dew point or condensation point. By moving more left taking away even more heat condensation begins to form. With enough time and heat taken away the water vapor turns to liquid. From here removing more heat and passing the Sigi point where water starts to freeze then taking even more heat away the liquid becomes solid.

From tested theories on atmospheric energy that it takes close to the same amount of energy to transform vapor to ice as ice to vapor. Our focus here is on droplets of ice on Grass not large solid ice blocks that require large amounts energy. Another advantage of Frost on Grass is that it is ice widely distributed in very small amounts.

We are interested in just enough heat added or removed to create and remove small droplets of ice or Frost. To create Frost this requires enough available water vapor and just the right amount of heat. We are in the Boundary layer where interactions are large and small at the same time. Most of the time and in some areas turbulent and unruly. Again our focus here is to develop simple heuristics that we can use to predict Frost events estimate the likelihood of a Frost event.

## Air and Gases

To look at Frost and Dew properly we start with air and the gases it contains. Regular atmospheric or normal air typically contain 78 percent Nitrogen 20 percent Oxygen and 1 percent other stuff. Remember the error margins as discussed above they will always come into play when predicting a Frost event. Just knowing the error margins are a big step toward determining the possibilities of a Frost event on the fly from one or more weather applications (app).

*Our focus here is on droplets of ice on Grass*

Air particles called bio-aerosols can be picked up dropped or carried for long distances. These are small particles that are picked up by wind like pollen, dust, some soils and dirt particle. Pollution and other combustible chemicals called aerosols are not easily broken apart by the Sun can be picked up and carried long distances as well. Now add relative humidity, wind speed, and temperatures there is a good start for the picture of what we are dealing with when estimating Frost.

Air does not behave like it looks. There are chemical interactions. Vibration interactions. Temperature and pressure fluctuations on immensely large scales they can be hard to predict and understand. Most of these are invisible or barely noticeable. All of these contribute to the complexity of determining a Frost event. Understanding the Frost event will also lead to understanding Frost recovery or when walking driving on grass does little or no damage.

Most Frost happens in still calm air where heat exchanges are undisturbed. Without wind mixing up and equalizing the temperature calm still air exchanges heat at a constant rate. Especially over grass. Calm still conditions warm the ground more effectively and also can release the warms as the night cools creating Dew. And if cold enough create a Frost event.

To get a full picture of the air at the time of frost or dew the best estimate is a controlled chaos. There are many different kinds of gasses, air particulates, soot, pollutants and influences from plants and trees and their growth cycles. Wind attempts to equalize temperature across a surface. Temperature triggers of the phases of water are the only thing we can really rely on for reference. The other is the influence of pressure on the phases of water which we will cover next.

## A multiple day No Frost Example

Over a three day span the first morning heavy Frost event followed by a cold cloudy day. Frost did not lift until late mid morning. Cold temperatures continued through out the day with a little warming yet temperatures did not get above 5C or 41F. A slight random wind non consistent. Clouds increasing. The second night cold cloudy and a dropping Dew point to below freezing. Night time temperatures hovered right above freezing in the weather app. A light Frost the second day. Warmed to just about the same temperature as the day before. A little more consistent wind. Frost lifted about an hour after sunrise. Clear and cold about the same as the two previous days. Still cloudy.

Lower clouds the third day with same temperatures freezing overnight Dew point within the error range and near freezing during the night. All the data points for a Frost event were checked. Calm morning no wind. Temperatures overnight close to freezing Dew point below freezing. Morning temperatures were cold. No Frost. Clouds moved out early morning with a strong directional wind. No Frost.

What happened? The storm that came through systematically removed the heat from the grass and the ground over the previous days. The cold nights and clouds kept the temperature cold and with just enough humidity to keep drawing the heat out of the ground. There was simply no more heat to be extracted so there was no condensation.

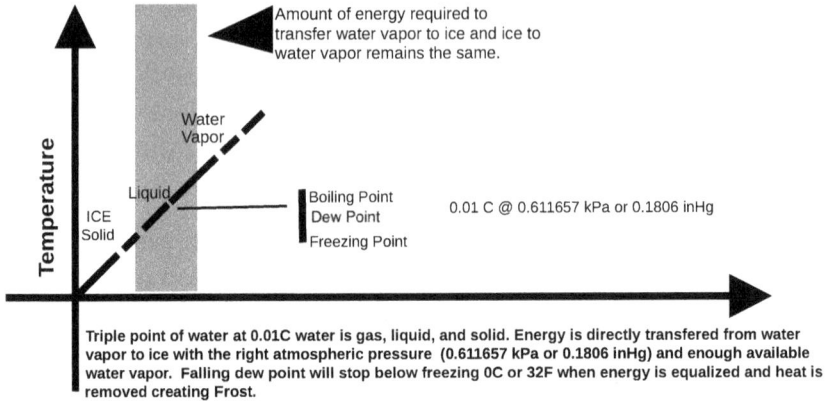

Figure 3: Triple point of water

Pressure can be thought of the weight of all the air above it on the ground surface. With the layer cake of atmosphere above now comes the influence of Boundary terrain. Mountains hills valleys and in some cases long cliffs and escarpments affect local weather systems. Wind can concentrate in valleys and be blocked and redirected by hills. Typically before, during and after storms where areas of low pressure cross to high pressure. Low pressure typically means turbulence and concentrated water vapor creating clouds and sometimes violent air movement or wind. A high pressure atmospheric system typically brings clearer calmer weather yet not always. The high pressure system is most of the time in relation to one or more low pressure systems. Most of the time clear days of and incoming high pressure follow storms and systems of a lower pressure.

The forces acting on the surface from a falling cooling air mass above represents a force pushing down when looking from below and a pulling down force from above. Or in the case of warm air a rising force pushing up and pulling up from below. These are attributes of the wild and free Boundary layer are some founding principles of atmospheric physics. The force of heat laden water vapor pulling up can be tremendous in a controlled environment. In an open system like the Boundary layer the rising of water vapor is an everyday occurrence and the rising air is filled underneath almost instantly from all directions.

Water vapor laden air is lighter than the surrounding air heats quickly and can rise quickly. It would be convenient to think this rising air has an impact. In an open system the rising air is simply replaced with air of a denser nature. Larger air masses are covered in the pressure section below. It is simply a swirling dance of equilibrium.

Like a cold drink on a hot Summer day the equilibrium wants to extract cold to equalize the temperature. Any water vapor in the air around the cold drink will drop its water right at the junction between hot and cold. This is the condensation phase change as mentioned above. One more step to get to a Frost event.

This is almost the same process as Dew on Grass except the air is cold and the grass is warm and radiating heat. This can be considered a phase change from water to ice.

The phase changes of water and its natural behavior is really the only reliable thing that we can lean on to explain Dew and Frost. Phase changes have distinct properties.

*Phase changes are all about energy equilibrium.*

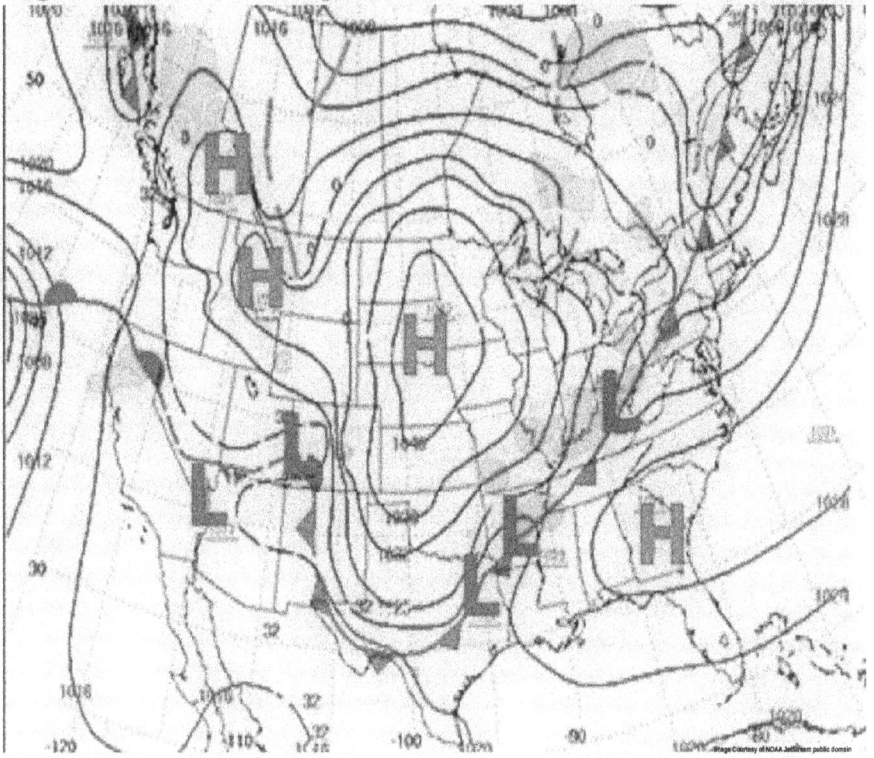

*Figure 4: Examples of high and low pressure systems. Several low pressure systems in a row and several high pressure systems. This is a picture of the effect of the Jet Stream affecting upper layers. This in turn affecting lower layers. Where is the Dew Point the lowest ?*

Fluctuations in potentials exist everywhere in the Boundary layer. Between gasses at various temperatures. Between the ground and the sky above. Between high pressure systems and low pressure systems. Energy takes the least resistant path to equalize. Water phases and the triple point of water get us even closer to understanding what causes Dew and Frost. Water and its behavior is what we are interested in and protecting grass from damage.

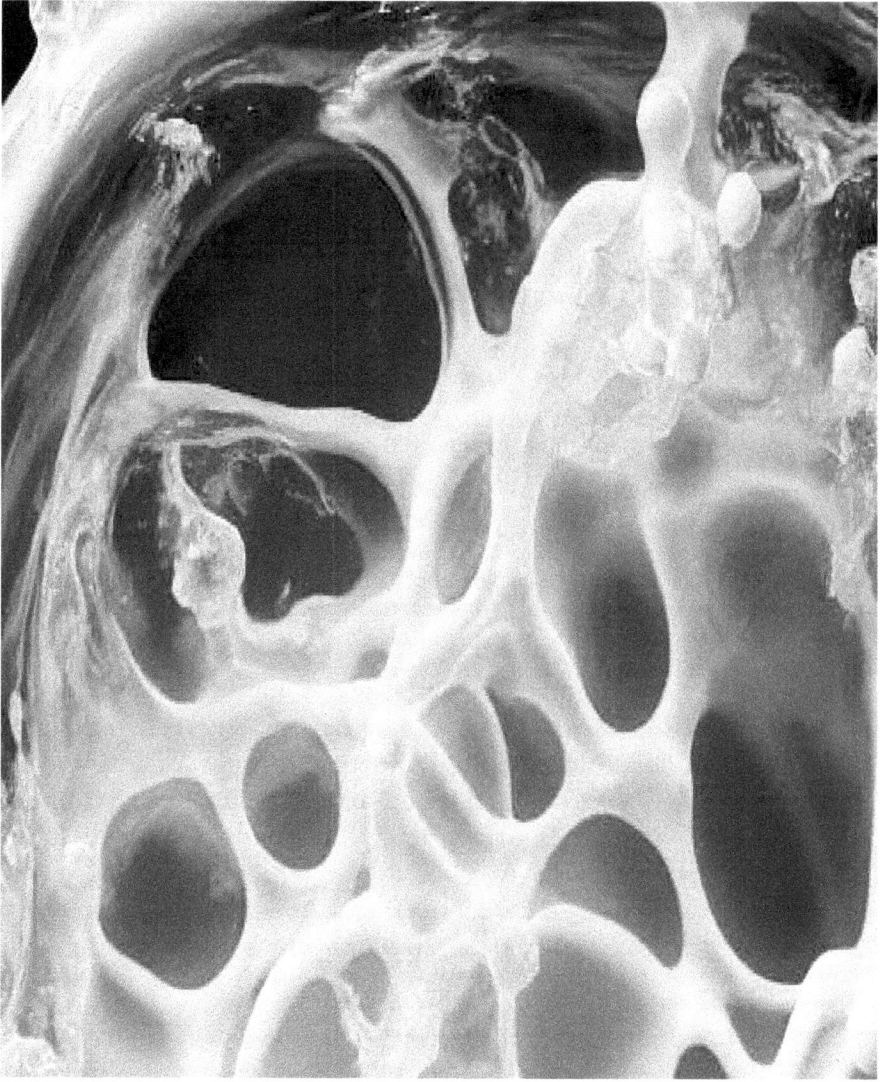

*Figure 5: Artificial Intelligence generated image of water vapor at the molecular level. It is constantly changing over time absorbing and releasing heat in various configurations patterns and vibrations.*

# Water Condensation point or Dew point

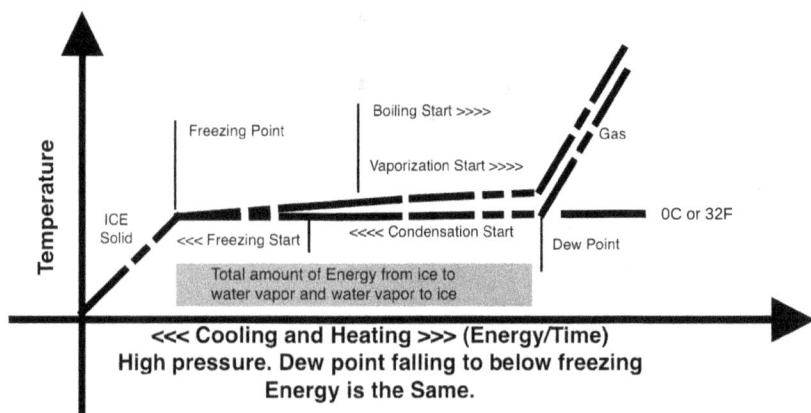

Figure 6: Bidirectional water phase diagram. Both dashed lines represent the transition of water phases at slightly different pressures as they approach freezing. When at freezing at the right pressure water moves freely through its three phases. Vapor to liquid to solid or Water vapor to water to ice.

Dew point is the temperature that water starts to condense out of air. This is air at its highest water carrying capacity. This temperature measure is dependent on the atmospheric pressure and the air temperature. To understand it better warmer air can hold more water vapor up to a point. The Dew point. If the same saturated air is cooled some of the water vapor will be forced out as water. If the same saturated air is subject to increasing in pressure water will be squeezed out. Decreasing or lifting the pressure on saturated air will increase the water vapor holding capacity.

It is important to remember that the Dew point is typically a calculated number based on relative humidity and temperature. Dew point is 100 percent % humidity.

*Dew point is expressed as the temperature air cannot hold any more water vapor.*

A little more on error margins. Adding up the error margin in Relative humidity 2%, the error margin in temperature plus or minus 0.01 C, add in some small processor division error and the error margin can be as high as 10-20%. Meaning a Dew point error of plus or minus of at least 4C or 5.4F max degrees is possible at the surface or ground level.

Frost happens when the real actual Dew point (temperature, humidity and pressure) number slowly drops at the surface level. So any one of these three variables can influence a Frost event. The easiest to predict is when the air temperature slowly drops to less than 0C or 32F or freezing. The temperature will continue to drop past the freezing point on most Frost events until it equalizes with the environment. To recap condensation is just about anytime the air temperature crosses the Dew point temperature. See the glass table example above page 14.

Frost happens when the temperature continues to fall past the freezing point of water 0C or 32F at a specific pressure. The pressure may simply be your height above sea level in calm air. The higher the altitude the less the pressure. Now add to this a large low pressure system which means typically a large area cold front or a high pressure or warm front. Pressure related frost can be exhibited by large Sun rollers that roll with the Sunrise in humid areas where rising and cooling air will roll across the terrain and at near freezing temperatures can cause an instant rolling Frost.

Error margins become really important when air temperatures get close to freezing. The error margins can indicate a possible Frost event on one extreme of the error and in reality it could go either way. Add in local terrain and pollution or soot availability presence of water vapor and heating of the ground and the Frost probability increases. When the morning comes no Frost. Temperature can be cold just above freezing 0C or 32F yet not cold enough.

Over a large area the weather app may say freezing yet in reality it may be 1C or 33.1 F on the ground. Add additional pressure from an incoming high pressure system and it is still not enough to create Frost. Or the humidity or water content in the air is just not there. Or the real existing Dew point at the surface is just not enough even though the weather app readings are the same as previous Frost events. These are some things to consider when dealing with conditions around a Frost event.

Figure 7 describes the general heat characteristics of a typical Frost event. An increasing freezing event from an in coming high pressure system can force the Dew point to drop below 0C or 32F. The falling Dew point is a great indication of Frost. Especially when the air temperature falls simultaneously with the Dew point. This is where it gets tricky. Without ground level warming from a previously sunny or warm day there may be little or no Frost.

Frost is a multiple day event. It involves a warming of the ground surface first. In our particular subject here we are dealing with grass which has up to 10 times the surface area of the ground underneath it.

All the little leaves of grass act like fins on a heat sink. As the air cools the comparison is like bringing a cold beer out of the cooler next to a hot and humid surface. Which is the same as a blanket of cold air on grass. There has to be some ground heating before a Frost event. If the ground is cold and no humidity grass may be frozen. This depends on the type of grass as cold season grasses have adapted and generate a kind of anti-freeze so they can use the typically bright sunshine on a cold clear day.

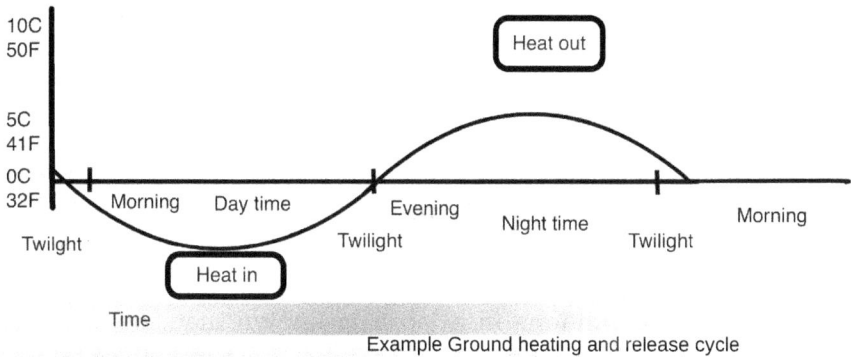

Example Ground heating and release cycle

*Figure 7: Ideal heat absorption under ideal soil and heating conditions. Real world varies considerably with Sun and heating exposure to the type of ground absorbing and releasing heat. Real world is way different. Presented here for reference.*

How much heating is enough. Enough to match the energy needed for condensing water vapor to water. About an 1 hour of intense sunlight at dusk. The ground has to have a difference. Wind, sprinklers, rain or other energy distributing events will disrupt this fragile mechanism. A calculated minimum amount of energy needs to be present at the time of condensation. Typically at night. Three or four hours of quality Sun to warm the Grass gives the grass some more time to transfer the warmth deep into the soil for later.

Like when warm breath creates condensation on glass or in the air on a cold tepid day. Or during twilight before the Sun comes up. When the breath is visible there is condensation. The exhale is the heating the air on a cooler to cold day forcing the water vapor to warm creating visible condensation. The airborne condensation created by the breath is surrounded by cooler to cold air changing the Dew point until it equalizes. A true demonstration of a phase change.

Wind will mess everything up. Wind is the great temperature equalizer and affect heat absorption and release characteristics. Wind will distribute heat. There may be Frost events triggered by wind would most likely happen hilly or steeper terrain where cold air is sitting on top of warmer air as the colder air may seep down the valley edges forcing the warm air up. This slight wind may cause a Frost if the conditions are right. Welcome to the Boundary layer.

Frost and Grass                    31

Calm clear skies are better for absorption and release. This is when most Frost happens. With a warm ground and the local air temperature cooling after dusk the temperature will eventually match the Dew point and Dew is created. Dew can be instant just like Frost. Like the cold drink the temperature differentials and the humidity has to be present for Dew to form.

We have to remember all we have here is the phase changes of water and a condensation point where air cannot hold any more water. This is the physical expression of Dew. In isolating atmospheric conditions in a lab it is known that condensation takes the most energy and time. The water to Frost transition takes slightly less energy than condensation.

Without the calm air there may be too much chaos for the energy exchange to make Dew or Frost. Wind is about the only thing that can disrupt a Frost. The Frost may still win if it is cold quick enough.

Figure 8 shows a Dew point crossing a temperature line creating a Dew event. Some modern weather apps can display a curve like this. Again there may be no Dew depending on the error and the terrain conditions. A heat in and heat out exchange is needed to start the Dew process. This process is the same as for Frost except the temperatures are lower and near or past freezing in the Frost event.

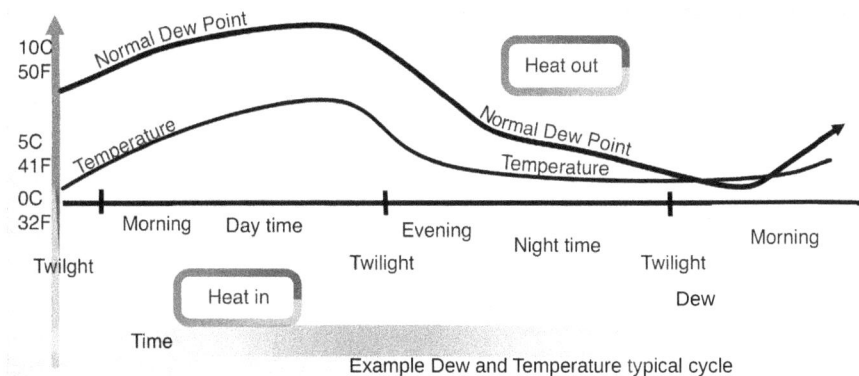

Example Dew and Temperature typical cycle

*Figure 8: A typical normal Dew or condensation cycle. Not all heat is extracted with Dew. Only enough to match the atmospheric circumstances. Dew is created when the Dew Point crosses the temperature line. Available on some weather applications as a micro climate feature.*

Dew point or condensation point lowers with increasing pressure. This graph shows an clearing sky with a possible high pressure front arriving just after sunset. This could also be the warm air rising from an area and a colder heavier air replacing the rising warm air. This is still a pressure difference. Open system variability can cause Dew by many different atmospheric situations.

The water vapor saturation point of water or Dew Point is variable by day depending on conditions. In figure 8 above and with some modern weather applications it is possible to see the Dew point drop over several hours. A significant rate of change of the Dew point is important to notice. A weather app may only display one or two incomplete drops in Dew point. Not to forget the error in calculations or the distance to the reporting weather station it may not always be accurate for your location. A simple example if there is a Frost event and all the numbers and graphs on the weather applications were way off and did not make sense it is important to take notes on where the numbers are and how far off they are for your area. The Frost event has a very tight set of conditions. Check the heat absorption from the day before even with cold weather grasses there could be some heat storage done by plants or other mechanisms.

*Condensation and Frost is a system of day and night events*

Both condensation and evaporation will help when estimating the level of condensation and if there is a Frost event. Of most interest is preventing damage to Grass. An example of this is Dew can freeze looking like glass. This is especially strange when dark or early in the morning light. This situation can be mistaken for unfrozen Dew causing damage.

*Sun to ground during the day. Ground to sky during the night.*

Frost takes two separate water phase changes. First condensation or water vapor to water and next freezing temperatures and ice or Frost. Water vapor to water is costly and requires more energy than the phase from water to ice. The triple point of water creates an easier transition between phases with the same amount of energy at less pressure. Condensation is important in Grass management. It makes for great mowing lines. It can help with estimating foliar feeding of grass. All of this related to water phases in the atmosphere.

The triple phase point of water is triggered just before air temperature reaches freezing if the pressure is right (0.006037 atm or 611.657Pa). For a reference on the pressure a place that is 9,940 meters or 32,600 feet below sea level without water. Water vapor can then go directly to ice with these conditions. This is an extreme example yet the Boundary layer is wild.

To summarize, error in estimating conditions with out real or actual feedback has been discussed. How estimating plays into the different dependant measurements like temperature and humidity. Phases of water has been touched on and the different requirements for Dew formation. The Boundary layer with all its quirks and caveats have been emphasized. Heat in and heat out equilibrium cycle over many days have been explained with respect to water phases, Dew point, Dew formation. A head first dive into Frost is next.

# Frost

Starting with the slow Frost formation the temperature drops slowly over the night. Conditions are still typically clear and calm with a slow moving high pressure system moving into the area. The heat from the day before slowly condenses to a mist above the ground or directly to water droplets. The slow change in temperature can be as simple as the day time temperature did not rise very much during the day. Or as complex as a high pressure system is slowly pushing in and back out again. Even a slight wind to distribute and release some of the stored ground level heat could stop condensation or dew before sunset. The main tell for slow Frost is if there is enough heat

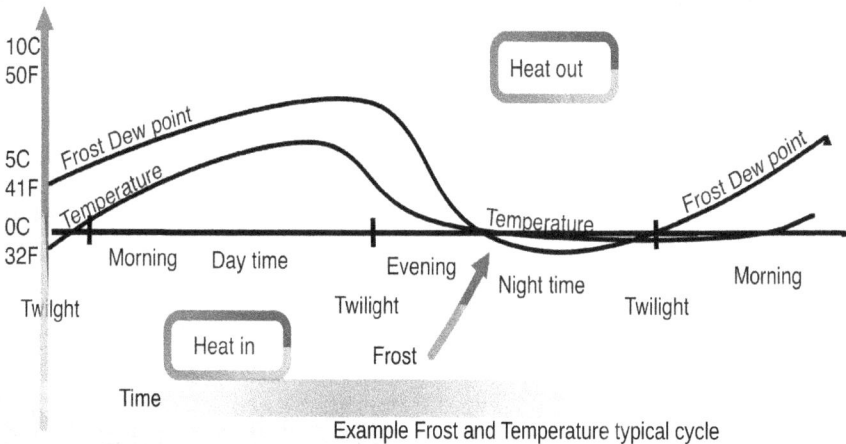

Example Frost and Temperature typical cycle

*Figure 9: A typical normal Frost cycle. Not all heat is extracted with Frost. Notice when the Dew Point crosses freezing Frost will form after a variable delay. If the Dew Point dips significantly below freezing the Frost will be instant.*

in the day before and humidity close to the surface on a clear and typically calm windless evening.

With slow cooling. Air temperatures drop slowly. When the air cools to below the temperature of the ground the energy exchange reverses. The ground starts releasing heat into the air. The water vapor is the first to absorb the heat and starts to rise. The air is cold. With temperatures continuing to fall. This limits the amount of water vapor that can be absorbed. This is a falling Dew point. The added effect is a water vapor saturated air mass descending on a slowly cooling ground. There is a slight increase in pressure from a slowly descending cold air mass on a cooling radiant surface. Eventually if the air temperature continues to drop and the Dew point falls below the temperature and the temperature is freezing a Frost event is likely. See figure 9.

The quick Frost event is like the slow frost even except with more potential heat in the day before. This also requires a clear and calm sky. There must be significant heating of the ground. A low pressure system like a rain storm followed by a high pressure clearing weather system that moves into the area in the evening or or morning before. The quick frost is where there is significant rain, a drying still and warming period, then a still and cold night. This creates two conditions that flag this as a possible quick Frost. One it creates sufficient humidity creating a possible thick ground fog. Two is a the significant heating may lead to the drop in Dew point as the temperature cools to below freezing at night. Each area needs to be considered unique as even with the same level terrain different areas can absorb and release heat differently. There are so many options in the Boundary layer that are dependent on local terrain and conditions. Incoming storm direction is not always the same as a high pressure incoming system. Low pressure systems can warm the ground with warm rain without Sun to ground during the day. Conditions can be just above freezing clear beautiful day and a Sun roller right at sunrise will roll a blanket of Frost through the whole area.

*Once the air temperature gets cold enough the condensation phase moves easily to the freezing phase; called deposition.*

Pressure events are the anomaly. Temperature and Dew point are simple matters when looking at pressure it can change everything. As the pressure decreases the boiling point drops. This is like boiling water over a mile above sea level when the water boils it is not as hot as it would be at sea level ( 1 atm or 101.32 kpa ).

With high pressure systems can produce pressure from 1 atm or 101.3 kPa to up to and around 1.072 atm or 1.08.6 kPa. Low pressure events start at sea level 1 atm or 101.3 kPa and drop to some where around 70 kPa or 0.70 atm. A dynamic play of invisible giants. In figure 9 at the start of the graph Dew point almost follows the temperature until the humidity is taken away after sunset. The Dew point which is a component of temperature and humidity is being forced into a steady drop past freezing and then stops with a bounce. Remember the Dew point is the point where water cannot hold any more water or air at 100% humidity. The components of the Dew point measurement are temperature and humidity. To reasonably say that humidity is taken away with respect to temperature creates this drop.

To explain with the frosty mug example the surface of the mug may be colder than the freezer temp as the water vapor is whisked away by the freezer and it is the surface temperature that is important as that is where the frost forms.

Clear still air is needed and a continuously dropping Dew point to slightly below freezing is a real tell for Frost. What this says is that any one of the three elements pressure, humidity or temperature may literally squeeze the heat out of the ground surface sometimes instantly leaving Frost as a trophy.

*Eventually all ground heat will be extracted as the season changes toward Winter.*

**Two glasses and a metal ribbon scrubby example**
*Take two glasses of the same type fill one with hot or boiling water put the metal ribbon scrubby in the hot water let it stand for say 10 minutes. Separate the scrubby and the heated glass and place all three items in the freezer. Close the freezer and Set a timer for thirty 10 minutes. Check on the Frost. Set timer for another 10 minutes and check again. Repeat until super frosty. Notice the amount of Frost on each one. The scrubby. The heated glass and the non heated glass. This is a graphic explanation of the natural Frost process. Frost on Grass is like the scrubby. The heated glass is like the sunshine heated wet soil. The room temperature glass is the reference.*

With enough stored heat in the ground the day before along with localized airborne water vapor present just before night fall the action of the radiant cooling from above will attempt to equalize temperatures from below. During the day the temperature above is equalizing with the air temperature. As day turns to night the heat is released into the atmosphere from grass and surrounding plants. This removal of heat from the surface condenses water just like the glass table example above. Water vapor holding heat from its phase transition rises displacing the air above it. This displacement may create a reverse pressure event. The temperature at surface level drops due heavy cold air pressing on the surface and the kinetic equalizing force takes over releasing heat. Leaving frost droplets at the surface level and cold temperatures close to the surface.

The more water vapor present in air the lighter air is. This is counter intuitive yet totally true and provable in atmospheric physics. Humid air is lighter or less dense than dryer air at the same temperature and pressure. See figure 5. If there is nothing but humid air in a large grassy area and the ground is releasing heat in the form of water vapor air close to the ground. As night falls the rising water vapor at nine meters per second 9 m/s can create a vacuum on the surface creating a pulling effect. Where there is a pulling effect the void created is filled with something again our system is open one thing affects another.

If the air temperature at the surface is near freezing or 0C or 32F the rising water vapor will remove heat from the surface making the surface colder than the surrounding air. The rising light air will force heavier colder air into the void causing an upward vacuum type pressure event.

Areas that are naturally below sea level have a larger column of air above them. The Boundary layer is full of these different areas where pressure and the natural play can create interesting and variable weather dependencies. An example, on the Western side of mountain ranges and small hills a small pressure event caused by the large air mass pushing against the hill can cause an increase in pressure which changes weather dynamics.

### A smudge pot example

There have been stories of Frosts moving over hills and down into valleys with villagers of old using smudge pots or smoke to save grapes and orchards from Frost hindering new growth or fruit blossoms. Again we have evidence of significant warming wet or watered ground typically in a valley with near by hills. These smudge scenes are typically in the evening just after dark on a still cool to cold night.

The warm heated vapor laden air rises from the valley floor pressing into a colder blanket of heavier blanket of air. This causes the heavier colder air to move in from the sides of the valley replacing the warm air. Once the colder air has blanketed the bottom the rising warm air mass ascends quickly creating a vacuum and the sudden replacement of that void with colder heavier air creates a sudden pressure event.

In this modern day the use of large fans is only effective to a point in stopping a Frost event. With what we know now without enough smudge pots the event would still happen. The heat laden smoke would rise with the water vapor laden lighter air almost helping the colder air move in underneath.

A light Frost might be alleviated with the added heat and particles from the smudge pots. Better to locate orchards and vineyards in an area where natural convection allows wind or pressure to naturally draw off or redirect the cooler air so it does not collect and create a pressure event.

There really may be something to growing on terraced hill sides in areas where there Frost is a problem. The natural Boundary layer may protect the plants by lightening the descending air mass with water vapor.

**Determining Frost Potential : First Signs**

☐ Recently wet or soaked ground.

☐ Sunny clearing day, no clouds, warm, calm, and no wind

☐ Significant warming of the ground and surroundings

☐ Clear evening and night with no wind

☐ High pressure system moving in

☐ Night time temperatures below 7C or 45F and falling

☐ Dew point falling.

☐ Temperature falling.

☐ Night time dew point falling to or below 0C or 32F

To use the above heuristic if one line is true mark as a one. Add up the the true statement and mark them as ones (1) and the not true statements as zero (0) . A score of 10 Frost is almost certain. Score of 8 or more Frost is highly predictable yet not certain. Less than 8 and more than 5 check local conditions. Less than 5 less likely. Always test ground temperature if in doubt. If the ground is still warmer than the Air temperature there is stored heat. A local weather station or handheld weather sensor, or an hand held IR thermometer can add data to a corrected prediction.

*Two distinct water phase changes water vapor to water and water to ice are used to create Frost.*

Some confusing situations like on some days there is no Dew is a great exercise to test your knowledge. This means there is no condensation. No presence of latent heat from the previous day or lack of water vapor to pull the water out of the air to lower the dew point. Wind will equalize temperatures. Clouds will trap heat. Water vapor will absorb heat and rise. This is typical of low humidity days with windy evenings or early mornings. Leaving cold with no Frost or no frozen ground.

## Frost Types

Frost can be light just like a light Dew or heavy from multiple freezing events literally squeezing water out of the sky turning it to Frost multiple times. Then the frozen surface event is also worth noting. Similar to an ice storm where the ground surface water is frozen in place causing a thin sheet of ice over everything is just like a Frost and looks different. We break these down to make it easier to communicate to others and to keep them labeled for simplicity.

**Evening Frost.** High pressure system moved out in the early morning. Significant early morning heating on a wet subsurface. Morning temperatures start above or near 7C or 45F and rising. No wind calm still day through the evening when another high pressure system moves in just before Sunset. A freezing event followed by a temperature drop leading to a Frost event just before dark. Night Frost. Calm and still morning and afternoon. Wet terrain. Temperature rising to well over 7C or 45F. Water vapor present near ground after cooling at dusk. Night time temperature drop to near 0C or 32F. With a freezing event sometime during the night with dropping dew point to 0C or 32F or below creating a Frost event. Stable high pressure over whole area. Area remains cold remainder of the night. If another low pressure system moves in before morning it may lift Frost by raining or adding snow to the event. Any warm wind at night above or near 7C or 45F may soften Frost event. Any addition of heat will help quell the Frost event. The heat is additive. So a warm wind for a short time will lessen the time for a Frost to dissipate after sunrise.

**Night Frost.** Several Frost events are possible during one night. The heat removed will nearly equal the heat put in during the previous day. Still calm air will exchange or remove heat with two separate events rather than one event. In a turbulent Boundary layer there could be two or more freeze events presented facilitated by the local terrain like a large depression or a leeward valley of a mountain range with a high pressure system moving over it. This could create one or more freezing events leading to removal of heat and Frost.

**Morning Frost.** High pressure system moved in during the late night. Calm still no wind. Clear with no clouds. Typical early Fall or late Spring event as ground needs to be warm and retain heat from a partially cloudy warm low pressure system the day before. The freezing event from the high pressure has to create a descending temperature to at least 7C or 45F or below. Dew point has to be falling and end up near 0C or 32F during the morning hours before sunrise. Tall rough may have Frost and nowhere else.

**After Sunrise.** An after sunrise Frost event is possible with the same conditions as above except the temperature has to be in the range of 6C or 41F dew point 0C or 32F or below. Add a freezing event such as a sun roller or $CO_2$ blanket slammer and bingo a Frost event. One has to see it to believe it. The other very cool weather event is the Sun roller. This starts as a typically clear and beautiful early sunrise in cold near freezing high humidity areas. A Sun roller looks like a haze just above the horizon with a clear sky above the observer. As the sun rises from level to the horizon to say one hand above the horizon the roller is almost at your location. To see a rolling Frost like this is impressive. The speculative cause of this kind of Frost is pressure from the rolling action of a water vapor laden rising air mass heated by the Sun forced back down by upper level colder air. This colder air falls yet it is the vacuum or pulling action left by the Sun side of the roller that creates the Frost according to refrigeration calculations where the leading side supplies the water vapor. This is a sight to behold. In and around the 35th parallel.

## Frost Events

A Cold duration really has nothing to do with the initial Frost. The real tell is how much radiant heat has been absorbed by the surrounding ground at the surface level the day before. The more energy absorbed the thicker the Frost. Frost is basically frozen water on a surface. This is directly related to the radiant heat absorbed during the day and then transformed as night falls on windless nights with the right temperature and Dew point.

The breath test. During frost events your breath will all of a sudden condensate as you exhale. Very similar to the condensation event your breath is warm the air is cold except this is a sudden change. When it is about to Frost the condensation of your breath will be the tell there is a possibility of Dew and if cold enough Frost. The way to tell is to check the surface level temperature. If at freezing or colder apply the heuristic.

There are a few more things worthy of mention that affect the formation of Frost. Pressure is one of them. There can be several different freezing events causing layers of Frost during the night. Clear calm cooler to cold nights enable the exchange of heat from the ground. If a daytime is cold or near freezing and nighttime is cloudy and below freezing for several hours grass maybe frozen. There is really no Frost. Grass can still freeze without frost. More on that in the grass section where we tie all of this and more together.

## Minimal Frost Requirements

**1.** First, Frost requires a typically clear or clearing sky with bright sunshine shining long enough to significantly warm any surface to a point that warms and can radiate heat out when it cools. Surface can be wet or drying as water will absorb heat.

**2.** Second, Is the ability of that surface to store and radiate heat quickly. This requires a typically windless condition yet very light and variable local wind maybe present. Forceful consistent wind will remove heat quickly especially on exposed areas and areas with a large exposed air volume like grass. Grass looks like an area ten times as big if the volume of all the grass leaves are counted.

**3.** Third, The a falling Dew point typically caused by one or more wide spread or local weather influencing events that create an opportunity . An incoming high pressure or local humid areas rising quickly because of water vapor accumulation or settling before sunset. For weather applications check when the Dew point crosses the temperature line.

**4.** Fourth there are always local exceptions that enable or disable a Frost event. Terrain that traps water vapor. Areas where wind is present even slightly or in wind protected areas. Local warming areas like forests that may trap and store heat overnight. The variations are almost infinite and a healthy skeptical attitude is needed and an ability to learn from events.

After Frost events exceptions are typically visible. And the evidence of error margin are evident. Sometimes the variable are consistent. Meaning Dew point is always below zero for your area for frost to occur when the temperature is near freezing may not be true for one area and true for all other areas. Other exceptions like weather station data timing delays and errors like really data taken at 6am with the time zone wrong making it an hour earlier or later. These errors are real and may influence a prediction introducing a difference between what is real and what is actual. More on this in the next section.

## Frost Dissipation

People are paid to protect grassy assets. These assets are valuable because the replacement cost is high and the loss of play may be many months. Leading to expense and loss of revenue.

Say you have players or a team that needs to use a grassy area. There is Frost event. How do you tell when the potential for damage has passed? From what we have learned here it could be as simple as using an IR thermometer and noticing when the grass is well above freezing and the air temperature is acceptable so that it does not freeze again. Some areas are extremely large and just pulling out large fans does not apply.

Watering expensive greens or other surfaces can be problematic early in the morning before sunrise on cold clear days. If the water temperature could be validated the amount of heat introduced could be calculated. From what we have learned here the timing of the water application should be taken seriously. When paid to protect an asset like expensive areas of grass it is good to have some tools to protect an asset.

Frost on grass is made up of small particles of frozen water on the leaves of grass. Leaves of grass are typically suspended away from the ground and surrounded by air. The Frost event has taken all the heat away until the grass thaws and starts moving its xylem to transfer heat from its roots. This is why the latent heat of the ground and the activity of the type of grass is important. If the grass is dormant it is still in sleep mode breathing and transpiring ever so slowly.

*Only one phase change is used in the thaw process : ice back to water*

A general cycle of recovery is when the Sun shines on grass it may trigger first the thaw or added heat to the leaves of the grass. Once this thaws this may trigger the transfer heat from below ground to the surface through the xylem or nutrient fluid to the leaves on the surface reducing thaw time. Cool weather grasses do this very well. Warm season grasses may be dormant or in the Winter sleep yet they are still living below the surface even though the exposed parts above the surface maybe just cellulose shells. The cellulose shells are typically reused so it is best to minimize or limit damage. That means driving or walking on it when it is still stiff and frozen. When the cellulose is pliable is will not fracture and break needing a new leaf in the Spring.

From the frosty mug experiments the time to thaw the mug takes as much time as to freeze it yet the frosty mug may dissipate its frostiness quite quickly yet the drink or glass is still cold for a while. How are you going to tell when the grass is ready for play? One consistent way is to check important areas with an IR thermometer to get surface readings well above freezing to make sure and use the back of the hand test.

As a reminder one water phase change to get frost or ice to water and another phase change to get water to water vapor. The condensation of water vapor to water is the most costly phase changes. Energy use for this reaction called the latent heat of vaporization is well known 2660 kJ/kg of energy to turn water to water vapor and 334 kJ/kg of energy for ice to water. This energy is bidirectional and it takes less time to get rid of the frost to water. Remember frost is created from water vapor moving to water then to ice. What this says is that Frost ice crystals may disappear quickly to water droplets with enough sunshine. This may be all that is needed and will be unique to your area and situation.

The ice may contain dust and pollutants and other particles that may influence or increase the thaw times. Local times may vary yet to thaw Frost on Grass can according to calculations take around an hour with increasing air temperatures from sunrise onward. Stiff frosty grass is the risk factor. Ice has to thaw to water to be safe on Grass. The Grass leaves need to be pliable and not stiff or frozen. Water on grass is relatively harmless yet the conditions should be well above freezing to prevent damage.

## Signs of Frost Recovery

☐ Sunny clearing warming early still calm

☐ Windy changing weather clouds moving out

☐ Rain or Drizzle warming slightly windy

☐ Low clouds or incoming clouds

☐ Low pressure system moving in slowly

☐ Dew point rising

☐ Temperature rising then steady

☐ Daytime temperatures above 7C or 45F warming.

☐ Not cold. Warming and nice.

To use this heuristic place a one beside each true statement add up the true statements. If the local temperature is still cold and overcast and there is no more heat in the ground Frost recovery will be slow. If there is fog and it is cold or high humidity above 80% then recovery will be slow. A score of 5 or greater can mean a faster Frost recovery time. A warm steady wind or a light rain can make Frost vanish instantly.

## Back of the hand test

Use the back of a bare hand and an infrared thermometer to help determine the temperature of the surface it should be above freezing by one or two degrees depending on the error margin on the thermometer.

To pass this test the back of the bare hand should be wet without ice crystals. It can still be cold and it is also wise to double check the grass leaves with a close look to check for stiffness or clear frozen leaves.

Yes error margins again. It is important because the Frost thaw event is complex. Some areas may still be frozen. It takes the same amount of energy to thaw a Frost as it does to make it and only a little bit of energy to dissipate the frosty ice to water.

Thawing a Frost will take less time when the day following a Frost is still clear and temperatures are warming up to and above 5C 41F for at least an hour. If there is any clouds they may delay the absorption of heat at the ground level. The way to get good at predicting Frost is to pay attention to Frost events and make notes on the recovery times for your area. Make notes of the areas where Frost lingers.

## Conclusions on Frost

Predicting Frost requires awareness of the day or days before the Frost event. The heat loading of the area of interest as well as the water vapor present the day before is critical to have for most Frost events. Calm clear conditions are also important. There are always exceptions in an open system like our atmosphere.

Frost formation starts with water vapor. This transitions to water forming Dew in the the first water phase transition to Frost. The second water phase transition is from water to ice forming Frost. Both of these phase transitions require very specific atmospheric requirements of temperature pressure and Dew point. The local environment such as geography, trees, cliffs and land forms like hills, valleys and flats that can affect Frost creation and thaw events and can delay and accelerate them.

Frost events form mostly during the evening, night and early mornings up and past sunrise. The amount of heat absorbed by the surface level on the day or days leading up to a Frost event contribute to the intensity which can vary from light to heavy. Other contributors are available water vapor and a still and calm arena for heat exchange through the late afternoon, night and on past sunrise.

Frost is frozen water crystals on a cold surface. The air is made up of not just known gasses and also contains pollutants, pollen, soot, and dust all with unique heat loading and off loading characteristics. These can influence Frost formation and corresponding phase transitions. Air temperature is critical for Frost 0C or 32F. The Dew point is the temperature that air cannot hold any more water and where the humidity is 100%. Both of these are influenced by pressure. Local pressure anomalies, high pressure systems, low pressure systems and wind can affect Frost formation.

Frost recovery requires one phase change where the creation uses two phase changes. Local water vapor creates the water first in the primary phase change for Frost. The secondary phase change transitions the water created in the first phase change into ice crystals. Phase transitions conserve energy meaning for each phase from water vapor to water and water to ice it takes the same amount of energy in both directions.

The type of grass is important. A cold weather grass has a type of anti freeze that enhances its ability to move heat from below ground to above ground. Other grasses are dormant in cold weather yet still breathing. Both have risks of damage when dealing with Frost.

To decide if a Frost event will occur a heuristic approach is employed in the section below. This heuristic uses specific items for each section. These sections are just a set of questions that are marked either true or false. The true answers are added up resulting in a score. This score is then rated. To answer each section on the early morning before a suspected event

These heuristics help determine the relative closeness of a Frost event. They are not perfect yet they get close. It is up to the practitioner who is on a journey to understand Frost to catalog anomalies and adjust them for their own local area. As locations for Frost vary widely around the world these heuristics are a good methodology for getting close to mastering Frost prediction.

An example, the heuristic is filled out early in the morning before sunrise including information from the night before. Remember error margins especially for wind if you are using an application weather data.

Day Before  = 5

Evening Before = 3

Night Before = 5

Morning of  ** Should already be Frost on the ground **

Add additional measurements with IR thermomotor to verify. In this example the falling Dew point happened just before sunrise.

**Frost event likely if all the sections are above a 5.**

## Frost Event Decision Tree

### Day Before
- ☐ Sunny and clear sky four or more hours before sunset.
- ☐ Long period more than four hours with slight or no wind.
- ☐ Significant heat absorbed by large surfaces. Warm ground.
- ☐ No wind at sunset.
- ☐ Water vapor present or Humidity > 55%

### Evening Before
- ☐ After sunset twilight Slight or no wind. Clear sky.
- ☐ Water vapor present. Humidity > 55%
- ☐ Air Temperature dropping
- ☐ Dew point falling and less than temperature
- ☐ Wet or drying ground

### Night Before
- ☐ At night Slight or no wind. Clear sky.
- ☐ Night time air temperature dropping
- ☐ Dew point dropping, near or below 0C or 32F
- ☐ Night Temperature 6C or 42F or below
- ☐ Ground surface temperature warmer than night temperature.

### Morning (of prediction)
- ☐ Before Sunrise Slight or no wind. Clear sky.
- ☐ Dew point slowly dropping to below 0C or 32F near or after Sunrise.
- ☐ Ground surface is wet and warm from day before.
- ☐ Ground temperature dropping
- ☐ Low pressure moving out and high pressure moving in
- ☐ High pressure moving in
- ☐ Water vapor present. Humidity > 55%
- ☐ Air temperature near or below 0C or 32F anytime
- ☐ Dew point less than temperature at any point during the night.

**Frost event likely if all the sections are above a 5.**
Supplement with additional actual ground temperature measurements. There is a possibility of one or more Frost events with enough ground warmth. If there is a Frost event and all sections are less than 5 check the criteria. These are general criteria and may not fit all circumstances adjust as you see fit. It is wise to make notes after an errant Frost event and adjust the decision tree parameters for next time.

# Grass

## Grass types and heat exchange

Grass and Frost have a unique dance. Living Grass respires or breathes all the time. Even dormant grass will very slowly respire and expel $CO_2$. When the sun shines non dormant living Grass will produce $O_2$ as a by product of making sugars for growth. These sugars are stored in the roots and combined with free soil hydrogen to extract and absorb soil nutrients. The xylem, sap, or nutrient fluid is carried from the roots to the leaves during growth for most grasses this typically happens at night. Sugar or fuel for this growth created and stored during the day. Along with nutrients xylem carries some heat to the leaves during the night and stores heat from the Sun naturally in the soil and roots during the day. In cool season grasses the xylem contains an antifreeze like chemical similar to car radiator fluid absorbing and releasing heat.

When grass freezes it becomes vulnerable to damage from breakage of the leaves. If left alone most grasses will recover from a good stiff Frost just fine. A hard freeze may penetrate the ground and freeze fairly deep into the ground. This is possible after several consecutive hard freeze events.

Dormant in a sense means that all the xylem or fluid in the leaves of the grass has moves down to the roots. The leaves of the grass are not growing and are merely cellulose shells or flexible reusable piping. This is a natural protection from the cold. Grass can then retrieve nutrients deep under ground in the roots for later use in the eventual Spring.

To explore grasses deeper there are typically two categories cold season grasses (C3) and warm season grasses (C4). Named after the initial carbon break down of Carbon Dioxide $CO_2$ into a three carbon structure called a C3 pathway. C3 grasses have a greater tolerance for Frost, lower light, temperature requirements and typically classify as fast growing cold tolerant or cool season grasses. These generate a special type of sugar similar to anti-freeze. This anti-freeze is able to tolerate some stiff cold events and not sustained cold to say -26C -15F. The roots will freeze with sustained cold or deep freeze past this temperature.

Warm season or C4 grasses refers to a four carbon structure designation. The C4 pathway breaks $CO_2$ into a four carbon structure. C3 pathway may still be present in C4 grasses. C4 grasses are more warm season grasses. C4 grasses go dormant in cold and very warm weather. Each cultivar of grass has its own weather or temperature triggers for protective dormancy. Some C4 grasses make an anti-freeze compound similar to the cool season grasses almost like a just in case survival strategy.

## Grass and Cold

When the atmosphere looks at the surface made entirely of grass it sees a surface of more than ten times (10x) what is really there. This is because the large amount of small fibers add up to a large area. This can be thought about as very large heat sink or heat source. Heat and cold can be absorbed up to ten times more than say a barren ground or a ground without grass. Grass does this naturally using the xylem as a heat distribution mechanism. Something to consider with a Frost event.

Different grass types react differently to cold and heat. With native grasses or grasses that have lived in a native environment for a long time they develop triggers based on the amount of light and temperature. An example is Zoysia a type of Bermuda grass it has rhizomes and stolons that was discovered in the northern latitudes. near Japan. Known for its long leaf and hearty resilience and the ability to adapt well to cutting soon became popular as a Golf course fairway grass. Originally from the Northern latitudes. Zoysia grass is sensitive to prolonged heat and cold is considered a warm season grass and will go dormant in both heat and cold. These genetic triggers are learned by the plant over eons. Growth can happen at night, during the day, on cloudy days, misty days all depending on grass type and its origins or the place where the grass has spent the most time throughout history. Remember grass has been interacting with the environment for eons or a very long time.

The key point here is that grass damage potential is based on grass type and time of year. If your area has many different types of grass at various stages of dormancy this could pose a problem for damage. If a grass is fully dormant then there is limited cause for damage if the subsurface is not saturated with water. Damage to the soil due to tire tracks can still impact roots of dormant grass during wet or soggy conditions in the Fall or Spring. If a grass is dormant and frozen there is still a chance that ice can fracture leaves and heavy vehicles can disturb the subsurface soil damaging the grass roots.

Typically one good solid freeze followed by consistently lower than normal temperatures is needed to throw grass into dormancy. The reverse is also true. To bring a dormant type of grass out of dormancy the temperature has to consistently be above freezing. There is some evidence that grass can also sense the length of the night as well as increasingly warmer longer days.

## Grass Damage

With respect to grass damage the Fall and the Spring are have the most potential for grass damage. Knowing the different types of grasses going into dormancy in the Fall can limit damage. Same in the Spring.

With Frost larger leaf grass types like tall rough have the potential to absorb and radiate even more heat than say a shorter grass like rough or green surrounds. Grasses like tall roughs have deep roots and can transfer heat from below ground to and from the ground surface. Long leaves acts like a radiator removing heat efficiently into the surrounding environment. Some Frost events just are in the taller grasses like tall rough for these reasons. Some types of tall grass radiate and absorb heat efficiently.

Depending on the grass cultivar and the energy movement from the roots to the base of the grass leaf different grass types will tolerate Frost differently.

If Grass is damaged by vehicles, foot steps, or other ways it will show up as dead or damaged grass. This will be evident with the tire tracks or the foot steps or other marks. To avoid this Frost is not always present under trees and these area can be used to park turn around or avoid damage if work needs to be done. If it is mandatory that work be done during a Frost event warm water may be used to remove the Frost before accessing an area. It does not have to be a lot of water just enough to warm the leaves of the grass removing the Frost. When it has warmed it might be beneficial to use plywood or other sturdy covering to access an area. Plywood will work well for dormant grass.

Using effluent water on frozen grass can backfire and create more ice if the temperature is cold enough. The analogy is the ice cream maker where salt and water are added to water in the proper ratio to freeze the outside of the container. The added salt and chemicals in effluent water have been known to create more ice when the intention was to get rid of a Frost. There will be a ground temperature range where this will be ok and varies by location soil and grass types.

Grass leaves can be damaged when parts of the plant are frozen and pressure is applied by heavy ice, boots, equipment or vehicles. Typically damaging the grass leaf, growth center or other plant parts breaking the cell walls and destroying circulation of the xylem. Cool season grasses may freeze or look frozen yet recover just fine as the special type of sugars and carbon compounds are resistant to freezing. Grass anti-freeze has a lower temperature limit. Roots can be damaged in Spring and Fall by disturbing the ground through compaction, tires walking paths or other soil disruption. Good roots make for good plants.

## Conclusions on Grass and Cold

The times with the highest risk for damage is early Fall and early Spring in the Northern hemisphere. Before the Fall and Spring equinox is typical start of Frost season. The days get shorter than the nights in the Fall and days are getting longer in Spring. This means more heating during the day and the potential for colder nights.

Grass stores and retrieves heat from below the surface. In a frozen state the grass leaves are vulnerable to breakage and the growth centers at the base of the leaves vulnerable to bursting while frozen or semi frozen. This will cause damage that can take many months to heal in the Spring as new leaves and roots have to be repaired or replaced taking energy.

To successfully manage cold or near frozen grass observation may not be enough. The history and source of the grass may need to be explored for more understanding. Also there is a possibility that a grass type may change or mutate after being in an area for a long time. In other words it may be sensitive to cold after new a planting and may adapt several years later.

Workers and players out in the Frost need to be educated on grass damage. With the information in here and some local experience consistent Frost prediction is possible. The estimated recovery or thaw time can be very much accessible. Most grasses can handle a little damage and recover quite well. Grass damage takes time and to fully heal may not start till the late Spring time. In the meantime those ugly tire marks or damage may be visible all Winter.

# Frost and Grass

To think other things like Grass is breathing around us at various rates sounds surreal. This is true and not so strange in reality. This breathing is not how we breathe. To listen to it for real would reveal a long slow rhythmic inhales when the Sun is shining and faster rhythmic exhales when growing at night.

Sometimes many different plants grass, trees, bushes are breathing in synchrony. All living plants breathe at various rates. This breathing exchanges atmospheric gases at different times based on what the grass has learned to adapt to over time. This breath in the form of gases are responsible for the Oxygen ($O_2$) we breathe. Grass will use the Carbon dioxide ($CO_2$) we exhale to make sugars. Combined with nutrients the xylem or nutrient fluid either delivers energy to the roots to be stored or delivers it to the growth centers or leaves for growth. This mechanism also stores heat and retrieves heat as well. Grass is producing both heat and large amounts of atmospheric gasses.

A great example, After a large grass fairway has been cut to length it will produce less Oxygen and exchange less heat.

Trees, grasses and soil microbes are a large contributors to Boundary layer gasses. A simple calculation on the amount of Oxygen $O_2$ produced by a Golf Course was 18 metric tonnes of Oxygen in one day. This is just a calculation for the Grass. For one tree 274 liters of Oxygen $O_2$ per day average can be produced.

Grass can produce up to three times the amount of oxygen than trees. The general grass production ratio is roughly 1:1 $O_2$ to $CO_2$. Add trees and other plants and this Oxygen $O_2$ number can almost double.

Carbon dioxide ($CO_2$) and heavier pollutants Nitrous Oxide ($NO_2$) can gather near the ground. Oxygen $O_2$ and related gasses can mix with water vapor absorb heat typically rise away from the heat source.

Calm clear nights are required for the energy exchange between the ground and the sky. The clear sky is one of the best Frost metrics. Heating the day before a Frost is almost necessary as well. Random potential pressure events confuse things.

In an open system there are many complex interactions. In large grass fields creating many tonnes of Oxygen O2 during the day may influence Frost formation. After night fall many tonnes of CO2 can be produced. Both of these gasses are components of air and invisible. We can reliably see a frost event and know what it takes to create one. The invisible involvement of open systems gasses may very well explain why Frost is created mostly on grass during clear calm nights.

The scientific information is just not there on pressure events. Although the author has personally seen and experienced a rolling Frost and a flash Frost anomalies the rowdy mystery has left me wanting. To personally see Frost cross a fairway in a direct line from the Sunrise and move directly West is an amazing site. And the whomp of the cold air influx before a lasting quick almost instant flash Frost is equally awe inspiring. There are pressure events. They work by increasing or decreasing the atmospheric pressure at the ground level surface altering everything mentioned.

On grass there is a little extra cushion and ten 10 times the surface area to remove heat. Heat then is directly absorbed by the grass like a radiator reversed. At night the reverse happens. Heat is radiated quickly heating local surface air in a cold environment the result is a quickly ( can be as fast as 9m/s ) rising air mass potentially leaving a vacuum. On flash Frosts there is a possibility that the vacuum could be strong enough to minimize the pressure to the triple point of water for just an instant which is all that is needed. This would follow modern refrigeration calculations. Again in an open system this may be more of a pointing finger and not really a proof or more like one piece of the puzzle not the whole picture.

In general when heat is removed Dew is formed and at the right temperature and pressure Frost is formed. On the contrary the same process to create Frost only takes a small subset of the energy to thaw the ice to water rendering grass damage nearly a mute point.

Frost In the early Spring and early Fall is more predicable due to the lack of massive amounts of Oxygen (O2) and Carbon dioxide (CO2) when trees have no leaves or grass is dormant makes the affects of. Dormant grass can still absorb heat and can still frost. With most dormant grasses the xylem is minimized into the roots and the grass leaves can still break.

## Limiting Grass Damage

There are protocols for all kinds of events. A process for dealing with an event not only frames defensive action as important is gives a group of people an appropriate response. For Frost the best action is to stay off frozen terrain. On or after a Frost event a Frost based protocol is a good approach for organizations, parks and playing fields.

To warn people the night before of a potential Frost may be good enough to avoid most grass damage. Even if there is no Frost event the following morning then a Frost protocol can be canceled.

A Frost protocol solves several problems. It lets people know that limiting grass damage is important. Also notifies people that there is a well known response that limits damage in the case of a Frost. Protecting grass may be as simple saying there is as protocol specifying to stay off grass when Frost is present.

The information on how Frost thaws becomes very important here allowing the safe estimation of an access time. This not only applies to the players it also gives the maintenance workers a set start time. If a golf course or other field needs to be prepared or cleaned in some way this will give an exact start time for the players or other general access.

In some geographic areas weather applications either lack coverage or weather station data may be unreliable or unavailable. Some useful hand held tools like an Infra-Red IR thermometer which measures a surface temperature remotely using an infra red beam gives an instant temperature of a surface. Hand held weather sensors are also relatively accurate. Complete with a small wind sensor, temperature and humidity sensors. Some even have a logging feature which can be useful. These are typically not replacements for a weather station yet on a budget may help make many people happy. With a little understanding of error margins the prediction of Frost can be turned into a fun exercise instead of an agonizing scenario full of demanding people depending on you to tell them when they can access a resource.

By using the lack of Frost under trees and knowing that on a sunny day an hour after Sun rise with temperatures over 6C or 42F for at least an hour will typically remove most Frost. This is another reason to have a Frost protocol; just in case there is a sudden Frost after sunrise.

## Weather applications

Error margins between real world temperatures and weather applications can be misleading when predicting a Frost event.

**When using a weather application for predicting Frost:**

☐ Find an accurate local weather source.

☐ Note its calibration and difference to local numbers.

☐ A local IR thermometer can check local temperatures and validate ground temperatures

☐ Portable weather instruments can add an additional reference point.

☐ Heuristics like the Frost decision tree can help with accuracy.

☐ Gather data the day before an estimated event. Remember heat

☐ Try a Frost protocol for communication.

☐ Frost parades demonstrate the potential of a Frost event to the crew.

☐ Satellites are used for large scale prediction and have trouble determining local ground level in variable terrain like golf courses. Refine your estimate with local ground level temperature checks.

There may be many unseen errors. The weather applications are mathematical model estimate of the real world. When in doubt check temperatures locally. Even stand alone weather stations can have errors. Year to year data points on a local weather stations may increase in error as equipment gets older. Or old equipment changed out for newer.

It is wise to check your weather equipment. According to the World Meteorological Organization reporting weather stations should be 1.25 - 2 meters above the ground and encased in a Stevenson screen or louvered box out of direct sunlight with adequate ventilation. As well as located in an open area away from obstructions like buildings trees and other interfering structures. Wind speed and direction sensors should be placed 10 meters above the ground in open areas. Rain or precipitation sensors should be at one meter above ground and also free from obstructions and rain direction interference.

Standard Temperature variance +-0.01 C

Standard Humidity variance +-0.2%

Standard Wind Speed variance +-0.5 m/s

Large format weather data must then sent to a central server in the right format. The data is then normalized and posted for use by the public. There are unseen and variable error each day from these normalizing calculations. There will be error. How much error is the question.

A magic variance number for Dew point can be as high as plus or minus 4C or 5.4F. This is still at 1.25 meters above the ground level. Which may mean significantly different air temperatures and Dew point at the ground level. There can also be low to the ground wind currents from local terrain that do not register in a normal weather station.

Pressure sensors as in atmospheric weather sensors are notoriously error prone and can go in and out of calibration. Atmospheric pressure measurement data variance over many different points should theoretically reduce the error yet with enough wobbly weather sensors can influence the difference between a predicted Frost event and no Frost event in your local area. A most interesting anomaly is the delay of the arrival of a high pressure system. In most cases pressure weather data get processed or smoothed and generalized to reduce error. In most cases the accuracy is generally useful. Its that one frosty day with a special tournament or outing that is delayed unnecessarily because of lack of preparation. It may not matter and be covered by a Frost protocol.

## Weather application example

The local weather data ups and downs as looked at through a weather application may be smoothed out in a big data set. Other local weather anomalies like an increase in atmospheric pressure caused by the squeeze of a valley that is pointed in the direction of a storm and the pressure increase from wind pressing into that valley. The rising of humid air during a still night and cold air rushing in to fill the void locally may not even be shown in a weather application.

There are services named "micro climate" that let a customer customize terrain and location. These are typically pretty good and really all that is needed for a lazy Frost prediction. When armed with a well known Frost protocol a level of confidence in the awareness of grass damage can be an low effort solution to grass damage by Frost.

With multiple year monitoring of an area there may be Frost windows in the Spring and Fall that may shift. The availability of past data can always be found yet determining a Frost window for your area may be more difficult without a weather application. For quality high end clubs members would appreciate the notice of an upcoming Frost window and a refresher on the Frost Protocol procedures.

In the calculated world of weather reporting where the amount of heat absorption by the surface is the toughest to estimate. The UV index may be useful. There is always taking shoes off and walking on a large heat soaked surface the day before a potential Frost event. As even devices that measure weather parameters rarely measure latent or absorbed heat.

In general there is a settling time with the temperature and the Dew Point for conditions to be just right for Frost. Sometimes this will happen at sunrise other times it will take hours for the high pressure system to settle and the lower temperature air close to the ground where it can do its thing a pull the heat out producing a Frost event. The important take away here is to determine the time of the Frost event and when it typically happens in your area.

# Early and Late Fall Potentials.

After the Fall equinox the days will begin to shorten and the nights lengthen. This means less heating during the days and colder longer nights. A low pressure rain storm followed by a high pressure system will only have so much time to heat the ground. With clear days calm days and bright Sun at lower latitudes below 45. After a good rain ground still wet and with a clear and warm day before sunset followed by a clear night with falling temperatures and a Dew point at or below freezing will give a good chance of a Frost event.

Thaw may take longer in the Fall with later Sunrises and longer nights. For a thaw the same amount of energy is required to thaw the Frost as create it. With a heavy Frost in the morning and say the morning temperature does not get above say 40 till noon. Even with a slight wind Frost may not melt to water for several hours after sunrise. With a warm wind from another direction the Frost may clear in less than an hour.

Most types of Grass that go dormant trigger on temperature before going dormant. Light in the Northern Hemisphere is getting shorter and opposite in the Southern hemisphere. Some grasses will minimize early Snow and Frost damage by sending some of the grass to sleep keeping some patches alive to make sugars late in the season.

After several significant freeze events grass will eventually go dormant. At this point all the xylem is out of the leaves and the plant is basically sleeping. Yet still breathing slowly if it is alive. A series of wake up events triggered on temperature can wake up dormant grass.

The back of the hand test can be used for verification that an area is frozen. Crushed or broken grass will recover in the next growing season if the roots are not frozen. Roots can freeze with a super freeze event where temperatures reach -27C or -15F for a sustained time without protection like snow.

## Winter Potentials

Snow may melt at various stages during the Winter months can be an obvious problem depending on environment. There may be areas that may be warmer under the snow. Damage to the roots must be avoided. Heavy equipment on partially frozen ground is a must mention. Although an obvious danger it is still worth mentioning.

Care must be used when removing tarps on golf course greens. As frozen grass may cling to the frozen tarps. Some places leave tarps on all winter. A tarp will absorb more heat since heat is trapped under the covering. Careful of warm daytime high temperatures as a tarp in a hot winter Sun can reach warm Spring time temperatures underneath. Frost will form on top of the tarp. Grass under tarps may still may be vulnerable when temperatures are near freezing. Sustained warm temperatures that may trigger a growth cycle. Good to measure green grass height in the Fall before Winter.

Blue light spectrum will pass through snow heating the ground underneath. Some hearty grasses may still store and retrieve heat energy from the roots in the winter. Some grass roots can grow well past 16cm or 6.3 inches deep. Heavy equipment or high traffic access needs to protect the Grass with some kind of weight distributing system like plywood. This will prevent damage and compaction to the soil and roots. Best to remove the snow layer if possible. Grass is a living community and it is still slowly breathing when dormant. Compacted snow can turn to ice. With weight can press into the Grass surface potentially damaging the strata.

Dormant grass can be cut out in large as possible pieces and replaced with care. The roots of the extracted grass should be protected with a tarp or moving blanket to minimize freezing damage to the root structure. Basically the anti-freeze generated by most grasses does not go down to much more below -27C or -15F. Meaning if temperatures are cold the Grass pieces may freeze.

## Early and Late Spring Potentials

After the Spring equinox swings the daylight length increases past even. The significant warming in the Spring season helps Grass come out of dormancy. New Grass growth may be tender and sensitive. There is more movement Random Spring cold events can delay and shock grass into many false starts using up stored energy reserves. For the most part grass will survive the ups and downs of Spring with some exceptions. If there is Frost there is potential for damage.

## Conclusions on Frost and Grass

The chaos of the Boundary layer is wild and free. With all its various mechanisms there exists a narrow window of circumstances that cause Frost. Grass adds to this by acting like a radiator by creating a surface area ten 10 times that of the soil underneath. In the Spring when Grass is actively growing. Always absorbing and radiating heat. Respiring and adjusting its breathing to the seasons as well as to temperature and sunlight. All this leaving a surplus of Oxygen O2 and Carbon Dioxide CO2 in the local environment.

The transfer of heat by Grass to the ground and back again through its on going respiration controls the heat release and absorption and acts like a buffer. Without stored heat presented to a cooler atmosphere of nighttime by Grass in the form of small leaves of radiant heat there is no Dew.

Without the right temperature and Dew point there is no Frost. The condensation event requires almost seven times more energy than the freezing event. And that is just an estimate. Condensation or Dew can be fast or slow depending on the pressure. The great reminder is a cold drink on a hot day when the condensation starts out slow and as the drink cools to a point the condensation stops. In the true topsy-turvy nature of the Boundary layer the air is colder than the Grass. The colder air mass is the cold drink and the heat is the Grass in our Dew and Frost situation. Dew forms right at the boundary of the Grass and air. There is no evidence that Dew or Frost falls to the ground. There are similar situations with ground fog and local collections of water vapor that can deposit as Dew. Frost is possible in the right circumstance. The only thing we really know is that to condense water takes more energy than to freeze water. To freeze water the air temperature has to be falling to below freezing. Frost is evidence that heat has been removed.

The only thing that seems to be a real tell for Frost is the descending Dew point and temperature slowly dropping to below freezing during the evening and nighttime hours. To predict this with modern weather applications without further local knowledge or input may be challenging.

The error potential on the Dew point calculations is notoriously bad and may look like it is going to Frost and does not. Measurements have to be almost extremely accurate to capture this which is way beyond the accuracy of most weather stations.

The Dew point dropping is a significant key indicator of Frost when the air is cold. For the Dew point to drop water vapor is being removed from the air by various atmospheric mechanisms. One possibility is the heated water vapor is slowly and consistently rising. Dryer cooler air replaces the rising air reducing the Dew point. When the Dew point and the temperature drop down to past freezing water vapor can go directly to ice without going through three phases. This will quickly equalize or release the latent heat and form Frost on the leaves of Grass.

Several of the other Frost events look like they form from increasing the air pressure from a pressure event. Pressure events are not officially recognized. To witness a flash Frost and Frost rolling across open fairways sure looks like a pressure event. These local events may act like a pressure event yet may be just higher level atmospheric air that is dry and cold being forced to the surface by rising water vapor or other displacement mechanism. This particular event is characterized by a rolling Frost in an East to West direction with the Sun angle barely above the horizon. Starting temperature around 5C or 41F on a clear cold morning right before sunrise at 6:54am. No Frost at this point. Then around 7:15am – 7:30am came the fast rolling Frost.

With in three or four minutes or less Frost everywhere. Then the whomp of instant freezing temperatures 0C or 32F at the ground surface. A heavy Frost on dormant grass no wind. Took about an hour and a half for the Frost to thaw or rise and turn to water or around 9:00am. The day before was clear and calm warm January day. It is good to keep detailed notes of anomalies for future use. Keeping track of both natural signs and measurements. A great option is to create an organizational Frost protocol.

What the above Frost event may look like from an weather application perspective would show clear and calm and temperatures a high pressure front moving in and a low pressure moving out. There would be no Frost warning. There would be no Dew point below freezing for half an hour then back up. It might pick up the rapid temperature drop. Yet the drop may not reach freezing. The data anomaly might get smoothed over by the algorithm and you would never see it.

The application air temperature error has been calculated may be up to 4C or 5.4F. This is taken as an error estimate of 100 weather sites with a temperature variance of 4-5C or 5.4 - 9F. Using different weather stations in various degrees of calibration add in the variations in humidity and Dew point calculations this is a reasonable amount of error. The error may not be always present and is a random error. So it will show up randomly.

Looking at Frost prediction from a modern phone app perspective the temperature may be accurate and drop to 0C or 32F and the Dew point may drop to 2C or 35.6 or 3C or 37.4 and stop. There is Frost. The reality is that the Dew point is already at freezing at ground level and the instrumentation or our current mathematical estimation of the real world does not match.

The solution in this case is to add additional data and observations. Keep notes and improve. Frost happens and if it happens deal with it attitude may not be enough to prevent damage. Put a frost protocol in place to prevent damage. Make it written and verbal and reinforce it with associates, players and personnel. A frost protocol can even be implemented the night before a prospective Frost event. What I have experienced is that knowledge is contagious and soon enough the club house or others will catch on.

Creating a Frost protocol with workers, club house managers, driving ranges and practice facilities is almost mandatory to maintaining a quality damage free environment. A protocol of this nature would communicate the back of the hand test for Frosty areas and implement set delayed start times. A creative manager would discount a breakfast or coffee for people betting that the Frost will lift early. Sometimes it might. This is the human value of Frost information. The key is to handle Frost events without damage. Efficiently and smoothly as possible. Can always do it the other way. If there is a Frost manage the reaction and the possible embarrassing damage till it has time to heal.

This book is really about people and how they can easily destroy a Frosty playing field or Golf course. The management of human resources around Frost can be costly. An example of this cost is having a 30 man crew wait several unnecessary hours because of lack of knowledge.

Frost damage is visible over a long period of time or in some cases over six months. Damage can be embarrassing and severe requiring sometimes immediate action like painting the mistake green or replacing large areas in the Spring time. The other reason for understanding Frost is the cost of the delay to playability.

A quality club or playing field should know when a Frost is going to lift. When to implement a Frost protocol. This is quality communication. Nothing more embarrassing than a golfer looking at a perfectly playable course when the club is closed because of Frost with no explanation or protocol. Or a practice field that can be played on safely yet the players and coaches are told no.

Understanding Frost for your area takes time. Being wrong about a Frost event with a Frost protocol in place is a real gift. Players get to play and understand the delay. A greater understanding of Frost is possible by changing your local criteria and conditions around Frost to improve the prediction. Damage to grass will be reduced if workers and others understand and follow a Frost protocol. Time will be saved by both players and workers with a little attention to detail around Frost recovery. Then there is the feeling of quality when a Frost event prediction is right on. And is totally unnoticed by players and management.

# Appendix 1 The air Psychronometric chart

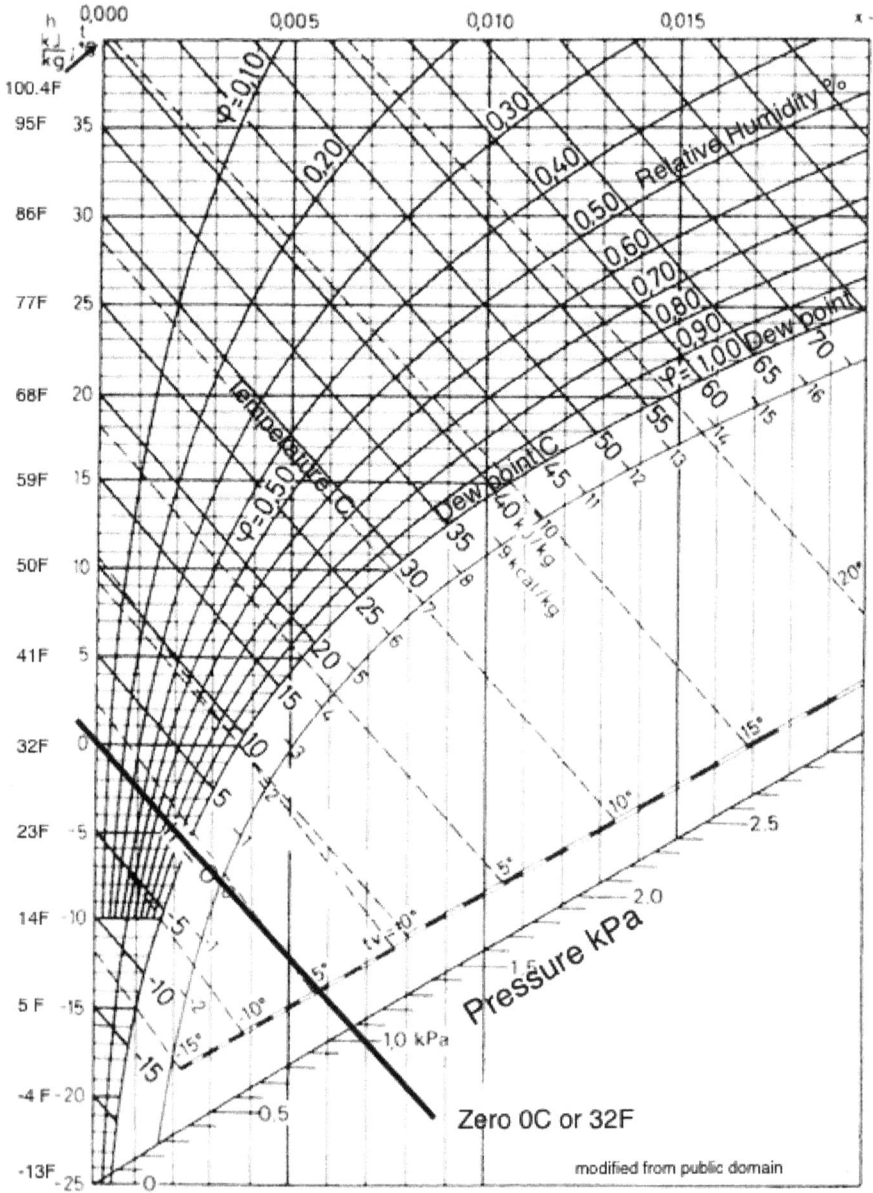

Figure 10: Air psychronometric chart for water. Shows water vapor relationships to air temperature and pressure at sea level. Can be used for as a reference in case instruments are suspect for error. The lines will not line up if a data point is way off.

The psychronometric chart is shows the relationship between temperature, relative humidity, absolute humidity or Dew point and atmospheric pressure. Temperature in C left x axis and pressure in kPa on the diagonal z axis. [(0.15psi = 1 kPa), (3.3863886667 kPa = inHg)] Humidity is the curved lines and is in percentages. The main curved line where humidity is equal to 1.00 is humidity at 100 percent % or the Dew Point. The curved lines radiating up are marked with percentages or Relative Humidity. Atmospheric pressure is on the diagonal and this graph is for barometric pressure at Sea Level. If your altitude is greater than Sea level there is less pressure environmental pressure.

To find A Dew point take a temperature on the left side say 10C at 0.60 or 60% RH follow it diagonally down to the right stopping at the 0.60 curved line. Now move across horizontally to the right and measure the temperature. Should be about 3.6 C. This is the normal Dew point for 10C at 60% RH. This is at Sea level. This number refers to a Sea level constant which in current understanding does not vary much with altitude.

To use the chart take a temperature measurement and a pressure measurement and follow temperature. Then find the pressure on the diagonal follow it up till it meets the temperature line and there is your estimate of the humidity.

Or take temperature and humidity and find the pressure. These are mathematically derived charts with implicit error margins. This is way to check if a system measurement is off or if your local area is experiencing something unique like a pressure event. The lines will not be consistent. Standard pressure at sea level is 101.3 kPa at 0 % humidity. Notice as well there is very little humidity below -10C.

## Appendix 2 Frost Protocol

A Frost protocol is about communication of the dangers and damage to Grass when frozen. The Frost protocol is an attempt to prevent a gross error like driving down a frozen fairway causing damage. It applies to both players and workers.

Protocol suggestions for staying off Frozen or Frosty Grass.

☐ Distribute the back of the hand test. If there is ice on the back of the hand wait at least 30 minutes till warmer test again and repeat.

☐ On a Frost delay post a mandatory start time no exceptions. Post signs. If there are instant Frosts in your area warn early patrons to look for signs. Delay play until the danger is over.

☐ Label the damage. Mr. X's damage. Jim's damage. Put a sign out identifying the grass damage. It can be there for months.

☐ Make it known with a small sign that damage will be visible even if sprayed green.

☐ Damage caused by Frost will heal the following or current Spring. Emphasize that Fall damage could be visible a long time in certain climates and zones. Even in the Spring after Winter.

☐ Distribute Frost protocol parameters in newsletters and member and employee communications or in on-boarding processes.

**Text for Signs** :

Frost Protocol : Stay off Frosty or Frozen Grass

Frost Protocol : Limit grass damage  Stay off Frosty Grass

Frost Protocol : Respect Frosty Grass Stay off.

Follow Frost Protocol : Stay off Frosty Grass.

# Appendix 3 Testing your knowledge

This section tests your real world knowledge of Frost and Grass. It gives a real world look into the importance of making learning about Frost and Grass a priority.

Pick the best answer. One answer per question.

**1.** In and around the Fall equinox day time air temperatures have been cold below 5C or 41F and cloudy low clouds for the last three days. You do not know if there has been any Frost or not. Time is 5am and you wonder if there is going to be a frost. Current temperature is 3C or 38F Dew point is 6C or 45F. What do you do?

 a) Call someone who might know. if there was a frost the last couple of days and ask them what they think?

 b) Check the phone application and check for yesterdays day time temperature and if the low pressure is moving out and when. And use a) for reference.

 c) Just implement the Frost protocol its too close to tell. It will not be warm enough for play till after 10am at the minimum.

 d) Make it known you do not care. If it Frosts deal with it. Make everyone come in early and sweep the shop. If you do not use the budget you will lose it.

**2.** Early Spring. Snow has gone. Daytime temperatures are 10C or 50F clear and calm. Time is 4pm they need an answer. There is a tournament scheduled for tomorrow and the club house and the teams demand a start time. The crew is new with no Frost experience.

a) Just say 10am and risk the fact that it will not Frost and be warm in the morning. If there is no Frost you will look like an idiot. Just take the flack. Your job is secure they are not going to replace you. Blow it all off. Tell them they are fools and where to put it.

b) Immediately check the Dew point and atmospheric pressure situation. Current temperature at 4pm is 4C or 41F and falling over the last hour. The low pressure system moved out earlier this morning. High pressure settling in. There is a slight breeze currently. No predicted wind. Clear calm skies predicted. You implement the Frost protocol. There is a clause in there that says no Frost by 9am and temperatures near 4C or 41F then all clear.

c) Take a tour and check the heat level with bare feet or an IR thermometer in several places. Check your notes from last year and you notice that the day length is quite short and cold the last few nights not freezing with light high clouds. You notice light high clouds with a slight breeze from the West. No predicted wind. No freezing temperatures predicted. You say 8:30am and turn the sprinklers on at 4am as insurance.

3.Winter no snow on the ground days are warmish in the 10C or 50F range. No Frost the last three days. With a cycle of cloudy rainy mornings and clearing warm beautiful days. A slow moving high pressure system is moving in mid day tomorrow. The Varsity Girls Team has an important tournament and they want to practice early and get on the road.

a) Tell them it will Frost tomorrow and implement Frost protocol. Too busy to deal with the attitude and all the demands.

b) You notice the high pressure is coming in early and tell the coach early in the morning that there is a high probability of a Frost event and to use the indoor facility.

c) You notice the high pressure system is coming in with some stiff wind that should be done early. It looks like a nice day no Frost. You mow and prep the course for their practice. They bake cookies and buy lunch for the crew after winning the championship.

4. Early Fall daily thunderstorms with 25C or 77F day time temperatures. Just before the equinox and the nights are cold not yet freezing. No Frost yet. A high pressure system moves in late in the day leaving calm clear skies dropping temperatures and dropping Dew point. Ground is warm and wet. Atmospheric pressure is high. Clear and calm.

a) You kindly tell your boss its going to Frost and the crew should have a bet and if he wins a Frost parade is necessary. He shuts you down rudely. He sneers turns and walks away.

b) You check the Dew point on your phone app and share with others in the crew the falling Dew point and temperature. Humidity is above 50 percent %. Include choice a).

c) A Frost event is likely

d) all of the above.

**5.** A new employee has just mowed the playing field and dropped his expensive mug on the field. Weather is cold and getting colder. After he finishes it is very close to freezing. The weather is cold and it is late at night there are no lights on. After cleaning his mower he takes a cart back out to fetch his mug. There are tire tracks across the field. He calls you on the phone and asks what to do.

a) Have him perform a hand test immediately while on the phone to see if there was ice on the back of his hand. There is a light Frost. So you tell him to go home and get some rest. The grass is Kentucky Blue Grass and it is a cool season Grass and handles cold very well.

b) Have him check the temperature of the Grass itself with an IR thermometer. It comes back at 0.5C or 33F. There is a visible light Frost he sends you a picture. You can clearly see the tracks on the frosty field. You ask him to take another temperature reading and it is now 0C or 32F and getting colder. You thank him and tell him to go home and get some sleep. It will be fine. The grass is a cool season Grass and as soon as the Sun comes up it will straighten right up.

c) You check the local weather station. The temperature is falling. The dew point is falling. The temperature is 0C or 32F and definite light Frost. You ask him when he finished and check the time on the weather histogram. You thank him and tell him it will be alright. The grass is a cool season grass and the weather station said when he picked up his mug it was just above freezing and had it been a half hour later it may have been a problem with a heavy frost.

d) all of the above.

1c 2c 3c 4d 5d

# Appendix 4 TL;DR

(Too Long ; Did not Read)

☐ Start with Appendix 3 and answer the questions the best you can.

☐ Review the glass table example.

☐ Check out figures 1, 2, 3, 4, 5, 10

☐ Open a cold drink or beer forget the can protector and just let the can or glass sweat. Or wet your hand and as you reach for the ice cubes for your drink and notice the water vapor coming off your hand. Mix the drink and watch the condensation on the glass.

☐ How long for the frostiness to disappear?

☐ Look at the water vapor picture again. Stare at it. Think about what you just saw vaporizing off your hand in the freezer.

☐ Perform the two glass and scrubby experiment. Frost on grass is on small leaves that radiate heat. Is this like the scrubby? How long did it take to Frost the glass? Did the scrubby take less time to frost? Compare the hot water glass to the room temperature glass is there a difference in Frost thickness?

☐ Read the short multiple day no frost example.

☐ Read the short section on weather application data.

☐ Take your phone magnifier and zoom in on your condensing can or glass. What is happening? Is the water vapor is visible?

☐ Take a kitchen probe type thermometer probe and place it in your drink. Ask yourself has this really been going on for eons with Frost. Ask yourself why you have not noticed it before. If you have trouble with it. Its invisible.

☐ Take the thermometer out of the drink and let it settle. Notice the air temperature. What is the difference? What does this have to do with keeping your drink cold?

☐ Take the quiz in Appendix 3 again. If you understand Frost stop. If not start again at the beginning of this appendix until you really understand the interaction between heat and a really Frosty glass.

## Alphabetical Index

www.ingramcontent.com/pod-product-compliance
Lightning Source LLC
Chambersburg PA
CBHW062104270326
41931CB00013B/3210